THE
Amateur Radio Dictionary

The most complete glossary of Ham
Radio terms ever compiled

Edited by Don Keith N4KC

EP
Erin Press
Indian Springs Village, Alabama

© 2015 by Don Keith

All rights reserved. No portion of this work may be reproduced without express written permission of the author except for brief excerpts in reviews. Please contact the author by e-mail at **don@donkeith.com** for permission to reproduce any portion of this book. This includes permission to reprint excerpts in such publications as Amateur Radio club newsletters or Ham Radio-oriented periodicals. These are encouraged and welcomed.

This work appeared in a slightly different form as a part of the book GET ON THE AIR…NOW!

Contact the publisher at the following address:

Erin Press
40 Red Stick Road
Indian Springs Village, AL 35124

Also by Don Keith

The Forever Season
Wizard of the Wind
The Rolling Thunder Stockcar Racing Series (with Kent Wright)
Final Bearing (with George Wallace)
Gallant Lady (with Ken Henry)
In the Course of Duty
The Bear: the Legendary Life of Coach Paul "Bear" Bryant
Final Patrol
The Ice Diaries (with Captain William R. Anderson)
War beneath the Waves
We Be Big (with Rick Burgess and Bill "Bubba" Bussey KJ4JJ)
Undersea Warrior
Firing Point (with George Wallace)
Riding the Shortwaves: Exploring the Magic of Amateur Radio
The Spin
The Road to Kingdom Come
The Ship that Wouldn't Die
THE Ham Radio Dictionary

Writing with Edie Hand:

The Last Christmas Ride
The Soldier's Ride
The Christmas Ride: the Miracle of the Lights

www.donkeith.com   www.n4kc.com

INTRODUCTION

Any pursuit that mankind might adopt quickly develops its own language, its proprietary jargon, its own semi-secret dialect fully understood only by its most avid and serious practitioners. Some even consider it to be a part of the hazing process, the price for becoming a member of the tribe.

Consider golf. Would a newcomer to the game know what a mulligan was? Or a divot? Would a novice fisherman be able to define a jig, a spinner bait, or a 30-pound-test line?

Amateur Radio is certainly no exception, though I prefer to think it is not really a hazing process. The hobby has now been around for more than a hundred years and has created through all that time its own dialect to be sure. Add to that the whole technical side and the governmental regulation aspect and it becomes even more of a challenge. All of this mumbo jumbo is not designed to flummox the newcomer at all. It is not some top-secret handshake to prevent others from entering a top-secret fraternity.

It is simply the way terminology develops and morphs over the years. Furthermore, with a pastime as dynamic and ever-changing as Ham Radio, it is inevitable that its verbiage changes and increases tremendously. Changes and increases so that even those who are active in the hobby may still need to be able to look up new terms as they come into common use. Or slip in while the Ham was busy in some other realm of this multi-faceted hobby.

Example: ask even the longest-tenured Amateur Radio operator to define "JT-65." Or to tell us what an "IOTA" is. Or to give a quick overview of "DSP" or "outgoing QSL bureau."

That, then, is the reason for this dictionary.

When I returned to the hobby after quite a few years of inactivity, I was amazed at how much everything had remained the same. But I was also amazed at how much it had changed. New technology, new operating events, new radiosport awards, new capabilities in our radios all made me wonder if I would ever catch up.

Many times during a QSO (see "QSO" in this dictionary) I wished for a quick-lookup that would tell me basically what it was that the guy on the air was talking about when he asked me a question or made a casual comment. Unfortunately, I often had to admit my ignorance and ask for clarification. Or hastily refer to Google, which inevitably sent me off into the weeds. Usually the fellow on the other end of the conversation was happy to explain it, giving me a chance to learn.

Still, I wished for some kind of Amateur Radio dictionary that would at least give me a basic idea of what the term meant and lead me to ask the right questions when someone dropped an unfamiliar word on me. As usual, I searched online for a glossary. Quite a few exist. Most are basic and woefully incomplete. Others have not been updated since the transistor. Some are totally technical and the definitions are not what I would expect to be helpful to someone new to all this Ham stuff.

I can only imagine how frustrating it must be for the true newcomer to our wonderful hobby. Or for others like me who might be returning to the hobby after a hiatus. Or for that guy in a chat on 6-meter SSB who is asked what he thinks of a "hex beam" or if he is in SMIRK.

What this dictionary is

What I have attempted to create here is a simple but more complete dictionary than has ever been attempted before—to my knowledge. It contains the most commonly encountered terms and jargon any of us might run across in Ham Radio.

I see the book as something that can be kept near the operating position in one's radio shack for quick access. Or have it downloaded on an e-book reader or tablet or computer for easy accessibility and search ability.

Language evolves. Even the brand-name English dictionaries are revised regularly to add words and terms that emerge naturally, or to cull those that are no longer commonly in use. Therefore I also see this work to be a continuing thing. That is where you can help.

If you would like to suggest a new word or term to add, or want to question my definition of a word, please email me at: **don@donkeith.com**. I will be happy to consider it for inclusion in this tome the next time it is updated.

What this dictionary is not

I will include a limited number of electronic or technological terms, those that the typical operator is most likely to encounter on the air or in group discussions with other Hams. Others are simply beyond the scope of this work and their inclusion might juts overwhelm the user. Besides, there are many other publications and web sites that do a fine job of that sort of thing, too.

First, I highly recommend the purchase of the American Radio Relay League's (see "ARRL") *The ARRL's Handbook for Radio Communications*. The index can lead you to easily understood descriptions and information about just about any technical term you will encounter in the hobby.

To purchase a new copy, or to find out more about Amateur Radio, visit the League's Web site at: **www.arrl.org**. You can usually find older copies of the handbook for sale at hamfests (see "hamfest"), in a boneyard (see "boneyard") or in a flea market (see "fleamarket"). They will only lack the very latest terms. Volts and ohms are still the same as they were a hundred years ago.

There are also many on-line sources for looking up electronics terms. They include, in no particular order:

http://www.maximintegrated.com/en/glossary/definitions.mvp/terms/all

http://www.csgnetwork.com/glossary.html

http://www.wilsonselectronics.net/dictionary.htm

http://www.hobbyprojects.com/dictionary/a.html

http://www.extron.com/technology/glossary.aspx

Again, this dictionary is designed as a quick lookup resource for newcomers (and old-timers, too!) to use when encountering a new or unfamiliar term while operating, listening, or reading. Here's an idea, though.

You may want to simply read through it at your leisure. Some term you thought you understood may mean something different altogether! Or you may happen upon some totally new ones. Of course, it may also spur you to suggest some that you feel need to be included, and I hope you will email those to me post haste.

Now, if you are new to Ham Radio, welcome to the world's greatest hobby. Enjoy! Learn!

If you have been licensed since the '20s, I hope you are still enjoying the hobby as much as you did the day you first fired up that spark gap transmitter (see "spark gap") and worked your first DX (see "DX"). But even you may find some new terms here that you are not sure about.

So 73 (see "73") and BCNU (see "BCNU") on the air! SK (see "SK").

Notes:

In addition to giving some pronunciations, I will also clarify what a number of terms are, using italics.

CW abbrev.: An abbreviation typically employed by operators using CW (Morse code), though you will often encounter some of them on other modes, including voice. CW abbreviations are quite often used on digital modes, making them even more important to know.

slang: Jargon you may hear while listening or communicating on the Amateur Radio bands. While many of these terms are in common usage and not truly slang in the hobby, I will still label them as slang if they mean something different in Amateur Radio than they do in the outside world.

In some cases, and to help you better understand the usage, I will give you origins of the word or term and examples of their typical use. I will also indicate those terms that are not totally accepted on the air, and even my opinion on whether that attitude is correct or not. In most cases, I would suggest you avoid the ones indicated as frowned-upon and either use plain English or my offered alternative.

antiquated term: a word, phrase, or term that is no longer in regular use but that you may encounter on the air at some point.

Please note that I mean nothing derogatory by use of the words "slang" or "antiquated." I have to label them as something!

Also note that each section begins with the NATO phonetic alphabet pronunciation (see "NATO phonetic alphabet") and the Morse code dots and dashes (see "Morse code") for the respective letter. I use "dahs" and "dits" instead of dashes and dots because the best way to learn CW (see "CW") is by sound, not as individual dots and dashes. A dash is a "dah." A dot is a "di" or "dit."

Be aware that there is a number-and-punctuation section after the letter "Z."

Now, from A to /, here is *THE Ham Radio Dictionary*.

A

Alpha ("AL – fah")

Di – dah

absorption - The loss of strength of radio frequency energy as it travels through any medium. In radio, this typically applies to the strength of a signal as it passes through portions of the atmosphere.

ABT (*CW abbrev.*) – About, approximately

AC - Alternating current. The flow of electricity in which the current periodically reverses the direction of its charge. See: **alternating current, DC, direct current**

ACC – Abbreviation for "accessory."

access code - 1. Term typically used when referring to a tone or series of tones used to access or activate a particular function of a repeater station such as a link, an auto-patch, or other capability. Numbers, letters, or other symbols are entered using a telephone key pad or properly equipped microphone.
2. A sub-audible tone on a transmitted signal that is required to access a repeater station. See: **CTCSS, repeater, tone**

A/D – Analog-to-digital. The conversion of an analog signal or source to digital. See: **analog, digital**

ADDR (*CW abbrev.*) – Address, as in mailing or shipping location

ADIF – Amateur Radio Data Interchange Format. A vendor-neutral personal computer file format intended to allow Amateur Radio station logs to be created in a variety of software programs so that they can more easily be used by other software.

adjacent channel interference – Interference to a receiver from a station operating on a nearby frequency or channel.

Advanced – A former class of Amateur Radio license. While many current Hams hold Advanced class licenses and they can be renewed, the Federal Communication Commission is no longer issuing them.

aerial (*antiquated term*) – Antenna. The term is considered outdated though still sometimes used in Europe. See: **antenna**

aeronautical mobile - An Amateur Radio station operating aboard an aircraft.

AF – Audio frequency, the range of frequencies that can be typically detected by the human ear, generally 20 cycles per second to 20,000 cycles per second. See: **audio frequency**

AFC – Automatic frequency control. A circuit in an electronic component that automatically compensates for drift in frequency. See: **automatic frequency control**

AF gain – In Ham Radio, usually refers to a control on a receiver that allows the operator to increase or decrease the amount of amplification that is applied to the audio signal from the receiver's audio circuit. See: **RF gain**

AFSK - Audio frequency shift keying. Using audio tones generated by a computer and transmitted over a voice mode such as SSB for digital communications as opposed to employing frequency shift keying (FSK) to vary the carrier frequency. See: **audio frequency shift keying, frequency shift keying, FSK, PSK31, RTTY**

after-burner (*slang*) – An external power amplifier.

AGL – Above ground level. The term is typically used when referring to towers or antennas.

AGN (*cw abbrev.*) – "Again."

AGC - Automatic gain control. A circuit in a receiver that maintains a more constant signal as it fades or increases in strength. The reaction time of the circuit to signal changes can often be varied. See: **automatic gain control**

A-index – A daily measure of the Earth's geomagnetic activity and how it is affected by solar activity. Generally a lower number means better propagation on the HF frequencies. A-index readings can vary from single digits to over 100. See: **HF, propagation**

airwaves (*antiquated term*) – The frequency bands on which radio communication might take place.

ALC - Automatic level control. A circuit in a transmitter's output amplifier that helps prevent amplifier overload or a similar circuit in an external amplifier that feeds a voltage back to the exciter to avoid overdriving the external amplifier. See: **automatic level control**

Alinco – A major manufacturer of Amateur Radio equipment, headquartered in Japan. Visit: **http://www.alinco.com/**

alligator (*slang*) – 1. A term applied to stations that run high power and have a transmit signal that can be heard farther than the operator's receiving capability allows him to hear incoming signals. Origin: Citizens Band radio, where such a station was said to be "all mouth and no ears." The term can also be applied to repeater stations that transmit farther than they can typically detect and re-transmit signals. See: **CB radio, Citizens Band radio, repeater**

2. A repeater station's time-out timer that shuts off the repeater transmitter if a user talks too long. Example: "You talked too much and the alligator got you." See: **time-out, timer**

allocation – Channels or frequency bands designated for use by specific radio services by a country's communications regulatory agency.

all time new one (*slang*) – A country, operator, zone, grid square or other entity which the station has never worked and/or confirmed a contact. Abbreviated as ATNO. Example: "Worked Indonesia last night for an all time new one."

alternating current - The flow of electricity in which the current periodically reverses from a positive to a negative charge and back. See: **AC, DC, direct current**

AM - 1. - Amplitude modulation. A type of modulation in which the amplitude of the carrier wave is varied in relation to audio when transmitted and then converted back to audio when received. See: **AM, amplitude modulation, carrier, carrier wave, FM, modulation**.
2. (*CW abbrev.*) – Morning.

Amateur– Typically refers to a person who is licensed to operate an Amateur Radio station by the government of his or her country. By law, Amateurs are not permitted to receive compensation for their activities, thus the "amateur" designation. Example: "Sue has been a licensed Amateur since 2009." See: **Amateur Radio, Amateur Radio Service, Ham, Ham Radio, pecuniary interest**

Amateur Electronic Supply – A multi-location vendor who sells Amateur Radio equipment and supplies. Abbreviated AES. Visit: **http://www.aesham.com/**

Amateur Extra – Currently the highest available class of Amateur Radio license in the USA. Usually referred to as "Extra class." See: **General, Technician**

AmateurLogic TV – A weekly television program for Amateur Radio operators, streamed live on the Internet. Archives of past shows are also available for viewing. Visit: **http://www.amateurlogic.tv/**

Amateur Radio - A non-commercial radio service established by international treaty that allows licensees to own and operate transmitting facilities on assigned bands. In the USA, the service is regulated under Part 97 of the Federal Communication Commission Rules and Regulations. See: **Amateur, Amateur Radio Service, Federal Communications Commission, Part 97**

Amateur Radio Data Interchange Format - A vendor-neutral personal computer-file format intended to allow Amateur Radio station logs to be created so it can easily be used by a variety of other software systems. See: **ADIF**

Amateur Radio Newsline – A service that provides news coverage of happenings in Amateur Radio and of interest to Hams. The news is disseminated on the ARN web site, on several Amateur Radio internet programs, as well as over many repeater stations around the country. Abbreviated as ARN. Visit: **http://www.arnewsline.org/**

Amateur Radio Service - A non-commercial radio service established by international treaty and regulated by each nation's government allowing citizens to be licensed to own and operate transmitting facilities using portions of the radio spectrum. The service was created for self-training, communication and technical innovation and to maintain a pool of operators and stations capable of assisting in the case of emergencies. The service requires that activities be carried out by licensed Amateurs solely with a personal aim and without pecuniary interest. In the United States, the service is regulated by the Federal Communications Commission under Part 97 of the FCC's rules. See: **Amateur Radio, Part 97, pecuniary interest**

Amateur Radio Supplies – A vendor of Amateur Radio equipment and supplies. Visit: **http://www.amateurradiosupplies.com/**

Amateur's Code – A suggested code of conduct for Ham operators, originally written in 1928. It reads:

The Radio Amateur is:

CONSIDERATE...He/She never knowingly operates in such a way as to lessen the pleasure of others.
LOYAL...He/She offers loyalty, encouragement and support to other Amateurs, local clubs, the IARU Radio Society in his/her country, through which Amateur Radio in his/her country is represented nationally and internationally.
PROGRESSIVE...He/She keeps his/her station up to date. It is well-built and efficient. His/Her operating practice is above reproach.
FRIENDLY...He/She operates slowly and patiently when requested; offers friendly advice and counsel to beginners; kind assistance, cooperation and consideration for the interests of others. These are the marks of the Amateur spirit.
BALANCED...Radio is a hobby, never interfering with duties owed to family, job, school or community.
PATRIOTIC...His/Her station and skills are always ready for service to country and community.

American Radio Relay League - The national association for Amateur Radio in the United States. Abbreviated as ARRL. The largest organization of radio amateurs in the world. Visit: **www.arrl.org**

ammeter – A test device for measuring electrical current. See: **amp, ampere, current**

amp – 1. - Ampere, a basic unit of measurement of electrical current. See: **ampere**.
2. (*slang*) – Amplifier. See: **linear, linear amplifier**

ampere - A basic unit of measurement of electrical current, typically defined as the measure of the electron flow through a circuit per unit of time. The term is often abbreviated as "amps" or with the capital letters "A" or "I". See: **amp**

amplifier - A circuit that increases the voltage, current, or power of a signal. The term often describes a device used to increase the output power of a transmitter or an internal circuit within a receiver to escalate the strength of a detected radio signal. Example: "I am using an amplifier here to boost my one-hundred watts output to about eight-hundred watts." See: **amp, linear, linear amplifier**

amplitude modulation – A method of placing information on a radio signal by varying the strength (or amplitude) of the signal's carrier in proportion to an audio signal. That variation in amplitude is then changed back to audio by a circuit inside a receiver. See: **AM, modulation, receiver**

AMSAT – The Radio Amateur Satellite Corporation. An educational organization whose goal is to foster Amateur Radio's participation in space communication. AMSAT is responsible for designing, building and placing into orbit Amateur Radio satellites. Visit: **www.amsat.org**

AMTOR - Amateur Teleprinting Over Radio. A digital communications mode in which error detection and correction are achieved by constant confirmation using "handshaking" or character repetition. AMTOR is still used by some Amateurs but has mostly been replaced by PSK31 and other more recent digital modes. See: **digital modes, PSK31**

analog – An electronic signal that carries information by varying time, spatial position, or voltage. See: **digital**

ANARC - Association of North American Radio Clubs. An organization of clubs primarily involved with shortwave radio listening. Web site: **www.anarc.org**

ancient modulation (*slang*) (*antiquated term*) – Derogatory term used for amplitude modulation (AM) by those who believe the mode causes unnecessary interference and uses too much bandwidth.

Anderson Power Poles – A commercially available type of power connectors used for quick 12-volt connections and disconnections. Visit: **http://www.andersonpower.com/products/singlepole-connectors.html**

angle of radiation – The angle at which a radio signal is emitted by an antenna and then refracted by the ionosphere. Lower angles of radiation generally result in transmissions that travel over greater distances. Higher angles may result in what is called near-vertical incidence skywave (NVIS) propagation. See: **critical angle, near-vertical incidence skywave, NVIS**

ANT (*CW abbrev.*) - Antenna. See: **antenna**

antenna - An electrical device or circuit designed to emit or receive electromagnetic radio waves. Antennas can take many forms.

antenna analyzer – A test instrument that enables checking various parameters of an antenna such as forward power, reflected power, standing wave ratio, impedance, and other factors. See: **forward power, impedance, reflected power, standing wave ratio, SWR**

antenna farm (*slang*) – Multiple antennas at a single station location.

antenna impedance – Resistance of a cable or antenna feed point in relation to the flow of electricity. Impedance is measured in ohms. Although an antenna's impedance fluctuates with the frequency of operation and many other factors, it should be approximately 50 ohms for most modern transceivers in order to achieve the maximum transfer of radio frequency energy from the transmitter to the antenna and into space. Most commercially manufactured coax cables for Ham use have an impedance of 50 ohms. The feed point of a typical dipole antenna is approximately 50 ohms. See: **antenna, coax, dipole**

antenna matching – Employing electrical components or devices to attempt to match the impedance of an antenna system to the output impedance of the transmitter and/or receiver in use. Matching helps assure the maximum transfer of radio-frequency energy from transmitter to antenna and out into space. See: **matching**

antenna modeling – Using computer software to create diagrams and charts of expected parameters and performance of an antenna. See: **NEC, EZNEC**

antenna party (*slang*) - A get-together of Hams to assist a fellow Amateur in putting up antennas or erecting towers.

antenna pattern – A diagram of the areas where a signal is expected to be stronger or weaker for a given antenna. Such a pattern can be plotted on a graph to help visualize where a signal should be best and worst. Each antenna design has a different expected pattern, but that can be altered in the real world by local obstructions, topography, propagation and other factors.

antenna switch – A device for choosing the connection of the output/input of a transmitter, receiver or transceiver when more than one antenna is available.

antenna tuner (*slang*) – A commonly used term for a device used to match the impedance of an antenna system to the output impedance of a transmitter or input impedance of a receiver. This term is considered slang because an "antenna tuner" does not "tune" the antenna at all. It merely attempts to find values of its components so that the impedance of the antenna system is close enough to that which the transceiver requires so the maximum transfer of energy to the antenna and into space can occur. See: **auto-tuner, internal tuner, match, matchbox, mismatch, transmatch**

antipode – Two locations at exact opposite points on the Earth's surface. The antipode of your location is as far in any direction on the planet as it can be from you. If you drilled straight from your location through the Earth, you would emerge at the antipode of where you started.

anti-VOX – Circuitry in a transmitter/transceiver that prevents audio from the receiver's speaker or noise in the background in the shack from actuating the VOX, a voice-operated relay that turns transmitting on and off when the operator is speaking. See: **VOX**

apogee – A point in the orbit of a satellite in which it is farthest away from the Earth. See: **perigee**

appliance operator (*slang*) - Hams who have little interest in building or experimenting with radio equipment. Instead, they are more likely to operate commercially available equipment, often with only a minimal understanding of how it actually works.

APRS – 1. Automatic Packet Reporting System. A system for real time digital exchange of information in the local area using Amateur Radio equipment interfaced with the Internet. The data may also be distributed globally for immediate access. This data might include messages, alerts, announcements, and bulletins, and can also be shown on a map display. Visit: **http://www.aprs.org/**
2. Automatic Position Reporting System. A capability of the Automatic Packet Reporting System in which GPS (Global Positioning System) information can be included in the shared data so the station can be tracked on maps on the Internet or on other properly equipped Amateur Radio devices.

AR (*CW abbrev.*) – "I have finished my message or transmission." Sent as one character: di-dah-di-dah-dit.

arc - An electrical flash between two conductors caused by the ionization of a vapor or gas.

ARC – The abbreviation for "amateur radio club." Example: BARC is Birmingham Amateur Radio Club.

Arduino - An open-source computer hardware and software company, project and user community that designs and manufactures kits for building digital devices and interactive objects. Many of these projects and programs can be applied to Amateur Radio uses. Visit: **http://www.arduino.cc**

ARES - Amateur Radio Emergency Service. A public service organization of the American Radio Relay League. Licensed Amateurs who have voluntarily registered their qualifications and equipment with their local ARES leadership, make themselves available for communications duty in the public service when disaster strikes. League membership is not necessary to participate in ARES but members do have to hold a valid Amateur Radio license. Visit: **http://www.arrl.org/ares**

ARISS – Amateur Radio on the International Space Station. Most crewmembers on the ISS are Amateurs licensed by their respective countries. Hams are able to chat with the astronauts during their down time and the ISS also schedules regular conversations via Amateur Radio with schools around the world. Visit: **http://www.ariss.org/**

armchair copy (*slang*) – Absolutely perfect copy of another station's transmissions.

ARN – Amateur Radio Newsline. See: **Amateur Radio Newsline**

array – An antenna system with more than one element.

ARRL - American Radio Relay League. See: **American Radio Relay League, League, The League**

AS – 1. (*CW abbrev.*) "Asia."
2. (*CW abbrev.*) "Wait for a moment, please." Sent as a single character: di-dah-di-di-dit.

ASCII - American National Standard Code for Information Interchange. A digital code for the transmission of teleprinter data. The ASCII 7-bit code represents 128 characters including 32 control characters.

ASR - Automatic send-receive. A radio-teletype (RTTY) terminal mode that allows message composition while simultaneously receiving text from another station. See: **radio-teletype, RTTY**

Associated Radio – A vendor of Amateur Radio equipment and supplies. Visit: **http://www.associatedradio.com/**

ATNO (*slang*) – Abbreviation for "All time new one." See: **all time new one**

ATT – Attenuator, attenuation. A device or circuit used to reduce the strength of a received signal, usually to block interference from a very strong and/or local station. See: **attenuator**

attenuator - A device or circuit used to reduce the strength of a received signal, usually to block interference from a very strong and/or local station. See: **ATT**

ATV - Amateur television. An operating mode in which an Amateur Radio station sends television signals over Ham frequencies. This term usually only applies to fast-scan television though Amateurs often employ a mode called slow-scan television. See: **fast-scan television, slow-can television**

audio frequency - The range of frequencies that can be typically detected by the human ear, generally 20 cycles per second to 20,000 cycles per second. See: **AF**

audio frequency shift keying - Using audio tones generated by a computer sound card and transmitted via a voice mode for digital communications as opposed to employing frequency shift keying (FSK) to vary the carrier frequency. See: **AFSK, FSK, frequency shift keying, PSK31, RTTY**

auroral propagation - Propagation of radio signals using highly ionized regions around the Earth's poles, often the Northern Lights (aurora borealis).

automatic frequency control - A circuit in an electronic component that automatically compensates for drift in frequency. Abbreviated as AFC.

automatic gain control - A circuit in a receiver that maintains a more constant signal as it fades or increases in strength. Abbreviated as AGC. See: **AGC**

automatic level control - A circuit in a transmitter's output amplifier that helps prevent amplifier overload or a similar circuit in an external amplifier that feeds a voltage back to the exciter to avoid overdriving the external amplifier. Abbreviated as ALC. See: **ALC**

automatic volume control - A circuit designed to keep a receiver's audio volume (loudness) at a constant level. Abbreviated as AVC. See: **AVC**

autopatch (*antiquated term*) - A device that interfaces an Amateur Radio repeater station to the telephone system. This allows a Ham using the repeater to make telephone calls over the air to any telephone. Sometimes simply referred to as a "patch," these devices have generally been replaced by cellular telephones. See: **patch**

auto-tuner – An antenna matching device that uses internal relays or other scheme to automatically search for the best match to the antenna system, as determined by a built-in computer rather than have the operator change the values of the device's components. The device does this automatically when it is engaged, usually with a button push or by sensing RF when the operator transmits. See: **antenna tuner, match, matchbox**

AVC - Automatic volume control - A circuit designed to keep a receiver's audio volume (loudness) at a constant level. See: **automatic volume control**

average power – The average power being run by a transmitter as measured on a standard power meter. The result is typically in watts. On some intermittent modes, such as single-sideband, the usual standard average-power meter is not physically capable of measuring the actual output power. This requires a peak-reading power meter. See: **peak-reading power meter**.

average power meter - The average power being run by a transmitter as measured on a standard power meter. The result is typically in watts. On some intermittent modes, such as single-sideband, the usual standard average-power meter is not capable of measuring the actual output power. This requires a peak-reading power meter. See: **peak-reading power meter**

AWG - American wire gauge. The standard for describing the diameter of wire. The wire size increases as the gauge number decreases.

Az/El - 1. The azimuth (horizontal) and the elevation (vertical) direction an antenna can be pointed.
2. A type of rotator that can change both the azimuth and elevation direction of an antenna. See: **rotator**

B

Bravo ("BRAH – Voe")

Dah – di – di – dit

B4 (*CW abbrev.*) – "Before."

backscatter – A form of propagation of radio waves in which those waves are reflected in the ionosphere back in the direction from which they originated. Signals so propagated can then be heard in areas they would normally skip over. See: **skip, skip zone**

balanced line - A feed line connected to an antenna that is made up of two conductors, each having equal but opposite voltages and with neither conductor at ground potential. See: **ladder line, open wire line, window line**

balanced modulator - A mixer circuit used in a single-sideband transmitter to combine a voice signal and the carrier signal, but causing the original carrier signal and half the audio signal to be suppressed. See: **single-sideband, SSB**

ball mount – A type of antenna mount with an adjustable swivel allowing an antenna to be mounted on a surface that is not horizontal or vertical. Usually used to mount an antenna on an automobile. See: **mobile**

balun – "Balanced/unbalanced." A transmission line transformer used to convert balanced input to unbalanced output or vice versa. A balun is typically used to couple a balanced antenna, such as a dipole, to an unbalanced feed line, such as coax cable, or to transition from a balanced line, such as open wire line, to an unbalanced line, such as coax cable. See: **balanced line, coax, dipole, ladder line, open wire line, window line**

band – A range of frequencies in the electromagnetic spectrum. Example: the 80-meter Amateur Radio band is between 3.5 megahertz and 4.0 megahertz.

band edge – The upper and lower limits of a band of frequencies on which an Amateur Radio station may operate. Operators should be careful to be sure no portion of their signals extend beyond the band edge.

band is changing, band is going out (*slang*) – Phrases used during a contact indicating that propagation conditions are changing and the other operator's signal is either fading or getting stronger. See: **fading, propagation, QSB**

bandpass – 1. The range of frequencies that might be allowed to pass through a filter or receiver circuit.
2. The range of frequencies that may be detected, heard, and/or displayed at any given time by a receiver.

bandpass filter - A circuit or component that passes signals in a defined range of frequencies while attenuating signals above and below that same defined range.

band plan – Frequencies that are reserved by gentlemen's agreement for specific types of operating, such as CW, DX, digital modes, AM, and more. Visit: http://www.bandplans.com/ See: **window**

bandwidth – 1. The frequency occupied by a particular type of radio transmission.
2. The amount of data that a circuit is capable of transferring.

barefoot (*slang*) - transmitting with the normal output power of the transmitter/transceiver and not employing a linear amplifier to further boost the transmit power. See: **amp, amplifier, linear, linear amplifier**

base loading – Typically the practice of using a coil located at the bottom of a vertical antenna to raise the inductance and give the antenna a lower resonant frequency. See: **center loading**

base station (*slang*) - A radio station that is designed to be operated from a fixed location instead of being portable or mobile (in a vehicle). See: **fixed station, mobile, portable**

battery - A device that converts chemical energy into electrical energy, storing the energy, and then makes it available as needed. Some batteries are re-chargeable.

baud - The number of distinct symbol changes (signaling events) made per second in a digitally modulated signal. See: **baud rate**

Baudot - A five-bit digital code, invented by Émile Baudot, and commonly employed in teletype/teleprinter applications. Pronounced "baw–DOH."

baud rate - The measure of the speed of data transfer for a modem. Pronounced "bawd." See: **baud**

bazooka – A type of dipole antenna that uses coaxial cable as its elements. The shield of the cable is the radiating element and the center conductor acts as a matching transformer to give the antenna a wider bandwidth than a typical dipole. See: **double bazooka**

BCI – Broadcast interference. Interference caused to other services by a radio station in the broadcast service, such as commercial AM or FM stations. See: **broadcast band**

BCNU (*CW abbrev.*) - "Be seeing you."

beacon – 1. A station that transmits constant signals, usually for the purpose of navigation, homing, or determining propagation conditions.
2. A light or strobe atop a tall tower to visually warn aircraft of the presence of the structure.

beam - an antenna that offers a directional beam pattern and usually featuring some rejection of signals from the back and sides as well as signal gain in the forward direction. See: **hex beam, gain, quad, Yagi**

bent dipole (*slang*) – A dipole but with one or both elements bent at angles to fit onto a smaller plot of land. See: **dipole**

Benton Harbor lunch box (*slang*) - A small, portable transceiver kit manufactured by The Heathkit Company, which was located in Benton Harbor, Michigan. Models were sold that offered users single band AM-only coverage on either the 10, 6, or 2 meter Amateur Radio bands. They were quite popular with Hams because of their price and portability.

Beverage antenna – A long-wire receiving antenna used primarily for 80 and 160 meters. Can be hundreds to thousands of feet long, is usually hung near the ground, and is terminated in a resistor to Earth ground at one end. Named for an early developer of the antenna, Harold Beverage.

BFO - Beat frequency oscillator. A circuit in a receiver designed to create an internal signal to mix with an incoming external signal in order to produce an audio tone for CW or to inject a carrier for SSB reception.

bicycle mobile - An Amateur Radio station operating a portable station while riding a bicycle.

big gun (*slang*) – A station with lots of high-powered, expensive equipment and antennas. See: **little pistol, peanut whistle**

bird (*slang*) – A commonly used word meaning "satellite." See: **AMSAT**.

Bird (meter) – Refers to a brand of directional wattmeter. "I'll bring by my Bird and we'll check your transmitter power output."

birdie (slang) - Spurious signals that are usually produced inside a receiver itself.

BK (*CW abbrev.*) – 1. Back ("Back to you.")
2. Break in ("May I break into your QSO?")
3. Break (End of transmission.). Sent as one character: dah-di-di-di-dah-di-dah.

bleeder - A large-value resistor connected in parallel with the output of a high-voltage power supply circuit in an effort to "bleed off" the stored current in the supply's filter capacitors once the supply has been turned off.

bleed-over - Interference from another Amateur Radio station operating on an adjacent frequency or channel.

block diagram - A drawing or chart that uses rectangles and other shapes to represent major sections of electronic circuits.

BN (*CW abbrev.*) – "Been" ("BN on air 2 hours.")

BNC - A quick connect/disconnect type of coax connector commonly used with VHF/UHF equipment,

boat anchor (*slang*) – An expression usually applied to older Amateur Radio equipment, primarily because the big transformers and other components used in such gear made it very big and heavy compared to modern rigs.

bonding – 1. Using highly conductive strapping to get a better electrical connection between station equipment, towers, antennas and Earth ground rods.

2. Using copper strapping to better establish low-resistance electrical connection between various parts of an automobile for a more effective mobile installation.

boneyard – A flea market or area at a hamfest where used Amateur Radio equipment, parts, and other items may be bought and sold. See: **flea market, hamfest**

boom – 1. The part of a beam antenna that runs perpendicular to the elements and holds them in place.

2. An adjustable structure that holds a microphone above the operating position so it does not take up desk space and the mic can be easily moved side-to-side or up and down.

boom set - Headphones with the addition of a small microphone on a boom so that it can be positioned near the lips. See: **headphones, headset**

bootlegger (*slang*) – Anyone who uses on the air an Amateur Radio call sign—whether it belongs to anyone else or not—that has not been assigned to him or her. A bootlegger often does not even hold a valid Amateur Radio license. Such activity is illegal and could result in fines and jail time.

bounce - reflecting a radio signal off of an object or other medium, such as bouncing a signal off the ionosphere, a water tank, the moon, or the tail of a comet.

BPL - Brass Pounders League. See: **Brass Pounders League**

bps - Bits per second.

brag macro, brag tape (*slang*) – A pre-set computer macro in digital mode software containing information about a station and the sending operator that is played back during a contact on digital modes.

braid – 1. The woven outer conductor of coaxial cables.
2. A woven (flat) conductor which gives a large conductive area and is often used for station grounding or for electrically tying together as many of an auto's metal parts as possible for mobile operation.
3. The highly-conductive woven screen of tiny wires around a wire conductor.

brass-pounder (*slang*) – A Ham Radio operator who sends Morse code (CW) using an older keying device such as a straight key. Sometimes used to designate any operator who enjoys CW. See: **straight key**

Brass Pounders League – An organization managed by the American Radio Relay League for Morse code operators who handle large amounts of formal message traffic through the National Traffic System. Abbreviated as BPL. See: **BPL, National Traffic System, NTS, traffic**

breadboard – Wiring up a proposed electronic circuit on a printed-circuit board, perf board, cardboard, or other such medium in order to test the concept and performance of the circuit before committing to a more permanent design.

break – 1. A term used to interrupt an ongoing conversation, especially on a repeater station. Such usage is typically discouraged. A station wishing to join a conversation should simply give a call sign. See: **breaker**

2. (*slang*) (*antiquated term*) – A term signifying the operator is ending a conversation with another station and starting a contact with another. Example: "Break with W8XYZ, and good morning, K4XXX." Common language is preferred.

3. In formal radio message handling, the term that indicates the preamble and sending instructions for the message are complete and the operator is going to now transmit the text of the message. The term is used again when the text of the message is complete and the operator is going to give the signature information. See: **BT**

breaker (*slang*) – An operator who desires to break into an ongoing conversation. Sometimes the operator will simply say, "Break," but this is not encouraged. The best way is for the operator to simply give his call sign. Example: "I'll stand by now for the breaker. Go ahead, breaker."

break-in – Employing circuitry when using Morse code (CW) to be able to receive signals between characters while transmitting. Full break-in enables an operator to listen to other signals between individual dots and dashes. Semi-break-in generally permits reception between characters. This allows the receiving station to interrupt the communication without waiting for the transmitting station to finish. This is sometimes referred to as "QSK," from the Q signal for "I can operate break-in." See: **full break-in, QSK, Q signal, semi-break-in**

brick (*slang*) – A small hand-held transceiver, so named because the original ones developed for the Amateur market were about the size of the typical construction brick. See: **HT, handie-talkie, walkie-talkie**

broadcast band – The parts of the radio-frequency spectrum assigned to commercial radio stations. In the USA, the AM broadcast band is 535 to 1705 kilohertz. The FM broadcast band is 87.9 to 107.9 megahertz. The channels between 87.9 and 91.9 megahertz are reserved for non-commercial and educational broadcasters.

broadcasting – Transmissions by radio or television that are intended for consumption by the general public at large. Broadcasting by Amateur Radio operators is not allowed except for bulletin dissemination and code practice by such stations as W1AW, the ARRL station. See: **W1AW**

BT (*CW abbrev.*) – "Break." 1. In formal radio message handling, the characters are sent to indicate that the preamble and sending instructions for the message are complete and the operator is going to now transmit the text of the message. The characters are sent again when the text of the message is complete and the operator is going to give the signature information. Sent as a single character: Dah-di-di-di-dah. See: **break**
2. Often sent during a CW conversation to indicate that the operator is moving from one thought or subject to another. Sent as a single character: Dah-di-di-di-dah.

BTR (*CW abbrev.*) – "Better." ("SIG BTR on this ANT.")

BTWN (*CW abbrev.*) – "Between."

bug (*slang*) – A semi-automatic key for sending Morse code that employs a spring lever to send a series of dots.

bumper mount – A device used to attach an antenna to the bumper of a vehicle.

bunny hunt (*slang*) – Using radio direction-finding equipment to locate a hidden transmitter. Such activities are usually conducted as a fun exercise or contest but skills learned can come in handy if an illegal, stolen, or malfunctioning, continuously keyed transmitter is detected and needs to be located. See: **fox hunting, RDF**

bureau – QSL bureau. Volunteer groups who help stations internationally to exchange QSL cards. They are typically maintained by a country's primary Amateur Radio organization, such as the American Radio Relay League in the USA. Stations keep the bureau stocked with self-addressed, stamped envelopes. When a DX operator works a number of stations, he or she will send a batch of cards to the bureau where they are sorted and sent on to the stations who have envelopes on file. For more on the ARRL's QSL bureau service for incoming cards, visit: **http://www.arrl.org/incoming-qsl-service**. See: **BURO, QSL, QSL card**

BURO (*CW abbrev.*) – QSL Bureau. See: **bureau, QSL, QSL card**

business communications - Any communication for the purpose of carrying on the regular business or commercial affairs of any party, something strictly forbidden on any of the Amateur Radio bands. See: **pecuniary interest**

busted call (*slang*) – An incorrectly logged call sign, typically in a contest.

C

Charlie

Dah – di – dah – dit

C (*CW abbrev.*) – "Yes," "Correct."

Cabrillo file – A computer-file-formatting scheme developed to assure radiosport operators would have a consistent way to electronically submit contest log data regardless of the software they or the contest sponsors used. Pronounced "cuh BREE oh." Visit: **http://wwrof.org/cabrillo/**

California kilowatt (*slang*) – A station that is running more output power than the legal limit

call book - a publication or CD ROM that lists all licensed Ham Radio operators by call sign and gives their mailing addresses as they appear on file with the Federal Communications Commission. These have been mostly replaced by web sites and logging software that interface with the FCC files. See: **logging software, QRZ.com**

call district – The designated and numbered areas of the USA in which the Federal Communications issues call signs with that district's number to any new operator who lives within its boundaries. Example: Any station licensed in California would have the number "6" as part of its call letters, such as K6ABC. However, at one point in time, if he or she relocated to another district, a new call sign would have to be requested.

Newly-licensed operators are still issued call signs with numbers that coincide with the FCC call districts but a station moving from one district to another no longer has to change call signs, nor do operators requesting vanity call signs have to request one with the district number in which they reside. See: **call sign, call letters, prefix, vanity call**

calling frequency - A frequency where, by gentlemen's agreement, stations may attempt to contact each other directly and then move to another frequency to continue the contact. Example: In the USA, 146.52 megahertz is the designated FM simplex calling frequency and no repeater station will have an input or output on that channel.

call letters, call sign – A unique sequence of letters and numbers assigned to and used to identify a licensed station transmitting a signal in the radio-frequency spectrum. In the USA, call signs are issued by the Federal Communications Commission. Amateur Radio call letters consist of a prefix of one or two letters, a call district number, and a one, two or three letter suffix following the number. See: **call district, prefix, suffix**

candy store (*slang*) - Term for a commercial Ham Radio equipment dealer.

cans (*slang*) – Headphones. See: **headphones, headset**

cap – Capacitor. See: **capacitor**

CAP - Civil Air Patrol, an organization that often calls on Amateurs to become members or assist in operations and drills.

capacitance - The measure of the amount of electrical charge that is held by a capacitor. Such a charge is measured in farads. See: **capacitor, condenser**

capacitor - An electronic component consisting of two or more conductive plates separated by an insulating material. A capacitor stores energy in an electric field. See: **capacitance, capacitor**

capacity hat - A system of wires or a solid metal disk attached to the top of a vertical antenna to counter its inductance, bring it closer to resonance, and increase its bandwidth.

capture effect – A phenomenon in FM transmission in which only the strongest signal can be heard even if others are transmitting on the same channel or frequency.

carbon microphone – A microphone design in which granules of carbon are used in the element to change audio to electrical energy.

carrier, carrier wave - A pure continuous radio signal with or without modulation. Such a wave can be modulated in various ways. See: **AM, FM, modulation**

carrier-operated relay - In its most common Amateur use, this is circuitry that senses a carrier on a repeater station's input frequency and causes the repeater to re-transmit that received signal. Abbreviated as **COR**.

CATV - Cable television. Delivery of television programming by cables that run into homes or other locations. In the early days of cable, the letters stood for community antenna television.

CATVI – Interference caused by or to cable television.

cavity - A very narrow filter that passes radio signals of a single frequency, usually used in repeater stations to protect the receiver from overload by a transmitter located on the same tower or nearby.

CB radio – The Citizens Radio Service, or Citizens Band. See: **Citizens Band, 11 meters**

CBR – Cross-band repeater. See: **cross-band repeater**

CC&Rs - Covenants, conditions, and restrictions. Refers to rules typically developed by homeowners' associations or real estate developers to maintain certain controls over what owners may do with their real estate properties. Generally, the goal of the CC&Rs is protect, preserve, and enhance property values in the community. However, they often restrict or prohibit a homeowner from erecting Ham antennas on his or her property.

center frequency – The frequency of an unmodulated FM signal.

center loading – The practice of placing a loading coil at the center of an antenna element in an attempt to achieve a lower resonant frequency for the antenna system.

CEPT agreement – Conference of Postal and Telecommunications Administrations (Europe). An understanding that allows Amateur Radio licensees from the USA to operate in most European countries without any further authorization or license requirement. This is different from reciprocal licensing agreements that exist with other countries. Visit: **http://www.arrl.org/cept** See: **reciprocal operating authority**

certification – Official technical approval by the Federal Communications Commission of electronic equipment intended to be sold in the USA. Except for power amplifiers, no Amateur Radio equipment requires FCC certification.

CFM (*CW abbrev.*) – "Confirm," "I confirm."

channel – A frequency or frequencies on which a station or repeater might operate. In Amateur Radio, only the 60 meter band has specific channels on which a station may transmit. Other channelization has developed as a result of gentlemen's agreements. Example: Channels are usually used in relation to FM repeater stations, and the frequency pair—such as "16/76"—is usually referred to as the repeater's channel.

charger – A device for restoring energy in a rechargeable battery. See: **drop-in charger**

chassis - A frame or housing for an electrical device such as a TV, transceiver, power supply, computer or similar equipment

chassis ground - A wire conductor that terminates on the chassis of a device for electrical grounding purposes. See: **earth ground, ground**

CheapHam – A vendor of Amateur Radio equipment and supplies. Visit: **http://www.cheapham.com/**

check in (*slang*) – 1. (*verb*) To announce one's presence and availability in a net and to list any traffic he or she may have to send. A station should only check in at the proper time to do so as designated by the net control station.
2. (*noun*) An operator who has announced to the net control station his or her presence and availability in a network as well as any traffic that he or she desires to pass.

chirp – Rapid changes in the carrier frequency of a CW transmitter, resulting in a chirping sound on the signal. See: **CW**

choke - An inductor used to block alternating current (AC) in an electrical circuit, while passing direct current (DC).

circuit breaker - A protective component that opens a circuit when an excessive current flow occurs. Similar to a fuse but a circuit breaker may be reset, not replaced, after the cause of the excessive current flow is corrected. See: **fuse**

Citizens Band - A radio service in the USA—and many other countries—often referred to as "CB radio." It is designed for personal and business use for short-distance radio communications between individuals on a selection of 40 channels within the 27 megahertz or 11 meter band. In the USA, no license is required to transmit on the CB channels. 11 meters was once a Ham band but it was taken away to create the CB service. Some resentment lingers. However, CB has been a common stepping stone for serious radio hobbyists to move on to Amateur Radio. See: **11 meters**, **CB radio**

CK (*CW abbrev.*) – 1. "Check."
2. The word count in the body of a formal radio message, referred to as the "check."

CL (*CW abbrev.*) – 1. "Call sign."
2. "Call." Example: "TNX for CL."
3. "Clear." Indicates that the operator is closing down his or her station. See: **clear**

claimed score – The final score achieved in an on-air contest (radiosport) as claimed by an entrant based on his submitted log. The log is subject to inspection and updating in most contests. See: **contest**, **radiosport**

clarifier - A control on a transceiver that allows the operator to vary the receive frequency a few kilohertz either side of the VFO frequency without affecting the transmitter frequency. Sometimes known as receiver incremental tuning. See: **RIT**

CLDY (*CW abbrev.*) - "Cloudy"

clean sweep (*slang*) – A term used in radiosport (contesting) meaning contacts have been made by a participant in all possible geographic regions available to work in the event. Example: Contacts made with stations in all ARRL sections is a clean sweep in the ARRL Sweepstakes contest. See: **ARRL, contest, radiosport, section**

clear (*slang*) – "Clear," meaning an operator is finished transmitting and intends to close down his station. Example: "K7XXX is now clear."

Clegg – A former manufacturer of Ham VHF and UHF equipment.

CLG (CW abbrev.) – "Calling."

clicks - Undesired "clicks" or "thumps" generated by a CW transmitter as the key contacts are closed or opened. Clicks can cause interference to other stations operating on the band. See: **key clicks**

clipping - Distortion that occurs when an amplifier is overdriven and attempts to deliver an output voltage or current beyond its maximum capability. Such distortion can not only degrade the transmitting station's signal but can cause interference to other operators.

closed – 1) A condition when a frequency band no longer supports radio propagation. Example: "Ten meters was closed after about 8:30 last night."
2) The operator has shut down his station.

closed repeater - A repeater station with access limited only to a select group. See: **open repeater**

cloud warmer (*slang*) - an antenna system which tends to radiate most of its transmitted energy straight up. This is typically not a desired trait. See: **worm burner**

Clover – An Amateur Radio digital communications mode that allows full duplex communications. See: **digital mode, full-duplex**

CLR (*CW abbrev.*) – "Clear." See: **clear**

Club Log - A free Internet site for DX enthusiasts that gives them the capability of producing DXCC league tables, DXpedition tools, log search services and most-wanted lists of countries. Visit: **www.clublog.org** See: **DX, DXpedition, DXCC**

club station – An Amateur Radio station that is specifically licensed to a Ham club, established and provided for use by its licensed members.

cluster (*slang*) – A web site on which stations report hearing ("spotting") or making contact with other stations. This allows operators who wish to talk to those stations to go to that frequency and attempt to make the contact. See: **DX cluster, DX spotting, spotting**

CME – Coronal mass ejection. See: **coronal mass ejection**

coax, coax cable, coaxial cable – An unbalanced wire cable with a center conductor surrounded by insulation and a braided-wire shield, all enclosed inside an insulating jacket. The shield is designed to reduce outside electrical and radio frequency interference. Impedance values of 50 ohms and 72 ohms are typical for coax in Amateur Radio use. Pronounced "KOH ax," "koh AX ee uhl."

code - 1. (*slang*) Morse code. A method of sending text information as a series of dots and dashes using on-off tones, lights, or clicks that can be directly understood by a skilled listener or observer. Today, and in Amateur Radio, the International Morse code is the standard version in use. The header of each letter in this dictionary contains the Morse version of the respective letter. Morse code was named for its inventor, Samuel F.B. Morse. In Amateur Radio, the mode is also often referred to as CW. See: **CW, International Morse code**

2. Digital bits and bytes in which each set is decoded as a letter, number or useful character. Example: baudot. See: **baudot**

3. One of several languages used in computer programming that may be interpreted as a set of instructions by a microprocessor.

code practice oscillator – A device that creates an audio tone that may be switched on and off using a Morse code key. Used to send practice CW. Abbreviated CPO.

Code proficiency run – An opportunity to copy Morse code during broadcasts from the ARRL station, W1AW. Correct copy submitted earns a code proficiency certificate for the speed copied. Visit: **http://www.arrl.org/code-proficiency-certificate** See: **ARRL, W1AW**

coil - A conductor that has been wound into a series of loops in order to increase the inductance in the circuit. See: **inductor**

color code – A system used to show the value of resistors or other electrical components. Numerical values are assigned to various colors that are painted onto the body of the components so the value may be determined. This is necessary because such components are often too small to have their values indicated by the actual numbers.

compression – Use of an electronic circuit that reduces the volume of loud sounds or amplifies quiet sounds by narrowing an audio signal's dynamic range. Compression is typically used by Hams to gain more loudness on their audio so they can be better heard or in an attempt to stand out in a crowd of callers. Using too much compression can result in distorted and harsh-sounding audio, which actually degrades intelligibility. See: **speech processor**

condenser (*antiquated term*) – Former term for a capacitor, an electronic component composed of two or more conductive plates separated by an insulating material. A capacitor stores energy in an electric field. See: **capacitance, capacitor**

conditions – 1. The state of the weather at the operator's location. Example: "The conditions here are cold and 28 degrees Fahrenheit."
2. Band propagation. Example: "Conditions are great this morning with signals strong from Europe."

3. The equipment being used by the operator, usually spoken as "operating conditions." Example: "My operating conditions are a Kenwood TS-590 and a dipole at 15 meters high." See: **CONDX**

CONDX (*CW abbrev.*) – 1. "Conditions." This can refer to weather, or band propagation. See: **conditions**

2. Station equipment, usually preceded by "working" or, on CW, WRKG)." Example: "WRKG CONDX HR TS-590 XCVR DIP ANT."

contact – A conversation on the air by two or more Amateur Radio operators. Example: "I enjoyed our contact last week."

contest – An on-air activity in which Ham Radio operators attempt to contact as many stations, counties, countries, zones, and/or grid squares as they can in a specified time period in competition with other Amateurs. Under the rules of each country, no cash prize or other award of value may be given for on-air contests. Winners typically receive certificates, plaques, or merely a listing in score results in publications or web sites. Visit: **http://www.hornucopia.com/contestcal/** See: **operating event, radiosport**

contest station – An Amateur Radio station built specifically for participating in radiosport on a grand competitive scale. Some stations are especially elaborate with multiple towers and antennas as well as operating positions for many people to man transceivers simultaneously. See: **radiosport**

Contest University – A series of seminars, webinars, and educational events for Amateur Radio operators interested in contesting/radiosport. Visit: **www.contestuniversity.com** See: **contest, radiosport**

control – A device that allows the user to vary the value or response of a component or circuit. Example: a pot may be used to control the audio gain of a receiver. See: **pot**

controller – The system or circuitry that controls a repeater station. This includes turning the repeater on and off by remote command, timing transmissions and turning off the transmitter if users talk too long, sending the station's call letters, controlling the auto patch, and programming and running the CTCSS encoder and decoder. See: **auto patch, CTCSS, repeater**

control link - The circuit used by a control operator to monitor and make adjustments to a station being operated under remote control. See: **control operator, control point, remote control**

control operator - An Amateur Radio operator designated by the licensee of a station to be responsible for the transmissions that are made by the station. Example: a club station. The station must be operated within the privileges granted by license to the control operator.

control point – Any location where a control operator oversees a station's operation including a station being operated by remote control. See: **remote control**

Coordinated Universal Time - The current time at 0-degrees longitude in Greenwich, England. This time is generally used by Amateur Radio operators when logging contacts in order to avoid confusion from time zones differences around the world. Abbreviated as UTC or GMT. See: **Greenwich Mean Time, Zulu time**

cop (*slang*) – Operators who take it upon themselves to fuss over the air at other stations who do not, in the opinion of the cops, follow proper operating procedures in a DX pile-up. Cops often cause more interference than the stations they chastise. See: **pileup**

copy (*slang*) – 1. To be able to hear another Ham Radio station's transmissions. Example: "I copy you just fine."
2. To receive a formal piece of message traffic.
3. To hear and understand Morse code.

copying (*slang*) "Listening." Example: "I was copying you and Joe last night on 40 meters."

COR – Carrier operated relay. See: **carrier operated relay**

corona ball - A round ball at the top of any antenna that would otherwise have a sharp point. The object is to minimize static discharge, which could damage the antenna or any equipment attached to it.

coronal mass ejection - A massive burst of gas and magnetic energy on the surface of the sun released into the solar wind. CMEs can affect radio propagation on Earth, depending on its strength and the angle at which it arrives at the Earth's surface. Abbreviated as CME.

counter – A test instrument used to measure digitally the frequency of a tone or signal.

counterpoise – Wires, metal elements or plates, or even an automobile body that forms all or a portion of the other half of a quarter-wavelength antenna. This helps make the antenna, in effect, one-half wavelength long and closer to resonance on the frequency for which it is cut. Wire counterpoises are often called "radials." See: **quarter-wave antenna, radials**

county hunter (*slang*) – A Ham Radio operator who attempts to establish contact with stations in as many counties and parishes in the USA as they can. Mobile stations often become rovers to allow county hunters to make contact and nets are established to enable county-hunting enthusiasts to pursue their goal. See: **net, rover**

courtesy beep, courtesy tone (*slang*) - An obvious audible sound on a repeater station indicating that a station using the repeater has ended his or her transmission. This tells other stations that they may now transmit. The courtesy beep also usually indicates that the talk out timer has been reset. Some operators add a courtesy tone to their own transmitter so other stations will know when they have stopped transmitting but this is not encouraged. See: **talk out, talk out timer**

coverage - The geographic area in which a repeater station provides relatively reliable communications. This can vary, of course, depending on the power, antenna, topography of the area, and elevation of the station attempting to use the repeater.

CPI, CPY (*CW abbrev.*) – "Copy." Example: "CPI U 599."

CPO - Code practice oscillator. See: **code practice oscillator**

CPS (*antiquated term*) - Cycles per second. This was once the way of describing frequency of any recurring event such as a radio signal. This terminology was replaced in 1960 by the term hertz. Example: "The transmitter was operating on a frequency of 7200 cycles per second." See: **cycles per second, hertz**

CQ (*CW abbrev.*) (*slang*) – A general call made by an operator inviting anyone to answer. The term may be used on all modes, including phone, CW and digital. A CQ may also be restrictive or directional. Example: "CQ Utah, CQ Utah. Looking for any station in the state of Utah for Worked All States." Or, "CQ DX. Looking for DX stations only."

CQ Magazine – A monthly magazine for Amateur Radio operators and other radio enthusiasts. Visit: **http://www.cq-amateur-radio.com/**

CQ zones – Geographical divisions of the world determined by *CQ Magazine* for the purpose of the operating awards and radiosport events sponsored and administered by the publication. For more, visit: **http://www.cq-amateur-radio.com/** See: *CQ Magazine*, **WAZ**

critical angle - The angle at which a radio signal is refracted by the ionosphere. Lower angles generally result in transmissions that travel over greater distances. See: **angle of radiation**

critical frequency - The highest frequency at which a radio wave will return from the ionosphere rather than passing right on through into outer space. See: **maximum usable frequency**

crimp connector – A type of wire connector that uses applied pressure to establish electrically and mechanically reliable connection rather than or in addition to soldering. A special crimp tool is typically used. See: **solder**

cross-band - Transmitting on one band while receiving on another.

cross-band repeat – Technology that allows an operator to extend the range of a low-power handheld radio by using his higher-power base station as his own personal repeater. The process requires both a dual-band mobile and base radio to retransmit on one frequency band a signal received on another frequency band, and vice versa.

cross-band repeater – A repeater station which has its input (receive) and output (transmit) on two different bands. Abbreviated as **CBR**.

CRT – A cathode-ray tube display, such as some computer monitors or television sets. This is rapidly becoming an antiquated term as CRTs are replaced by flat-panel monitors.

crystal - A piezoelectric device designed to resonate at a particular frequency. Most crystal frequencies or frequency ranges depend on the material from which the device is made, its dimensions, and the temperature at which it operates. Temperature extremes can cause a crystal to resonate at an undesired frequency, different from the one for which it was designed. See: **crystal filter, crystal oscillator, quartz crystal**

crystal filter - A network of crystals used to obtain high rejection of unwanted signals outside the range of the filter's designed operating frequency range. See: **crystal, crystal oscillator**

crystal oscillator – An electrical circuit that employs a quartz crystal to maintain as accurate and constant as possible the frequency of a transmitter or receiver. See: **crystal, crystal filter**

crystal set – A very basic radio receiver made with simple parts and a wire antenna. It requires no external power source to work, only the radio frequency energy generated by the transmitting station. Visit: **https://www.midnightscience.com/**

CSCE - Certificate of Successful Completion of Examination. This is a document certifying that a person has successfully passed one or more elements of the Amateur Radio license examinations in the USA and will not be required to retake the test on that material. Visit: **http://www.arrl.org/exam-element-credit**

CTCSS - Abbreviation for the term "continuous tone-controlled squelch system." This system uses a series of sub-audible tones on a transmitted signal to restrict access to some repeater systems. Unless the proper tone is present, the repeater will not re-transmit the signal. Such codes are used to keep distant signals from causing the repeater to re-transmit or to overcome noise or adjacent-channel interference. They are also employed to keep a repeater station closed except for a designated set of users. See: **access code, closed repeater, tone**

CU (*CW abbrev.*) "See you."

CubeSat – A small, cube-shaped Amateur Radio satellite designed for launch into Earth orbit. See: **AMSAT**

cubical quad – A vertical wire antenna formed as a loop with four sides of equal dimensions. Quads usually are made up of two or more elements. Because of size, they are mostly used on 20-meters and up. See: **quad**

CUD (*CW abbrev.*) – "Could."

CUL (*CW abbrev.*) - "See you later."

current - the flow of electrons in an electrical circuit. See: **AC, alternating current, DC, direct current**

cut numbers (*slang*) - A way of sending numbers in Morse code (CW) by substituting shorter letter characters for the longer number characters. Typically used in contests to speed up the contact rate. Example: Instead of a report of "599," the operator would send "5NN." The number Ø is often sent simply as "dah."

cutoff frequency - The frequency at which a filter will begin to reject signals that fall outside its designed operating range.

CUZ (*CW abbrev.*) – "Because."

CW – 1. Continuous wave. This is an unmodulated, uninterrupted radio-frequency wave.
2. (*slang*) Morse code emissions or messages, even though they are an interrupted wave, broken to form CW characters. See: **code, International Morse code, Morse code**

CW Skimmer – A software program designed to decode Morse code transmissions that can be heard within the passband of a receiver. With some software-defined radios, this can be an entire band or more worth of signals. Such information is now being voluntarily submitted by operators and made available for general viewing on a web site. Visit: **http://www.dxatlas.com/CwSkimmer/**

cycles per second (*antiquated term*) – the number of complete cycles completed each second in an alternating current or radio-frequency signal. This term was replaced in 1960 by "hertz." Abbreviated as CPS. See: **CPS, hertz**

D

Delta

Dah – di – dit

data communications – The transfer of data between two or more locations.

Dayton – Usually refers to the world's largest Amateur Radio get-together, the annual Dayton Hamvention in Dayton, Ohio. Visit: **http://hamvention.org/**

dB – Abbreviation for decibel. See: **decibel**

dBd – Comparison in decibels of the gain of an antenna to a theoretically ideal dipole antenna, one in a vacuum in free space with no interaction with the earth beneath it or any other potential obstructions.

dBi – Comparison in decibels of the gain of an antenna to a theoretically ideal isotropic (uniform in all directions) antenna, a point source of electromagnetic or sound waves which radiates the same intensity of radiation in all directions.

DC – Direct current. The flow of electricity in which the current flows in only one direction. See: **AC, alternating current, direct current**

DE (*CW abbrev.*) – "From," or "This is." Example: "WA4GIY DE N4KC."

deceptive signal – Any transmission by an operator, whether properly licensed or not, that is intended to mislead or confuse. Example: Someone reporting an emergency when none exists. Such transmissions are illegal in the Amateur Radio Service. See: **false signal**

decibel - A unit used to express the ratio between two values of a physical quantity one of which is typically a reference value. Often used to indicate the strength of sound or RF waves. Abbreviated as db. Pronounced "DESS uh bul."

delta loop - A variation of the loop antenna in which its continuous wire "legs" form a triangle shape. See: **loop**

demodulation - To extract the original information-bearing signal from a modulated carrier wave or signal. See: **modulation**

desense – A reduction in receiver sensitivity due to overload from a strong, nearby transmitter.

detector - The stage in a receiver in which the modulation information is recovered from a radio-frequency carrier wave.

detection – The process of converting in a receiver circuit a radio-frequency signal into a form that can be further processed by the receiving device, such as from RF to audio.

deviation – During the modulation of FM signals, the maximum amount that the frequency changes on either side of the original carrier frequency. Deviation that is too low can result in the audio being difficult to hear. Too high deviation can cause clipping of the transmitting station's audio or interference to adjacent channels.

dielectric - A non-conductive material used to separate two conductors such as the foam or plastic between the center conductor and shield in coaxial cable. Air can be used as a dielectric in such components as capacitors.

DIFF (*CW abbrev.*) – "Difference."

DigiPan – Digital Panoramic Tuning. A freeware software program that offers the capability for a panoramic display of PSK31 and PSK63 signals over a broad frequency spectrum on a computer screen. Visit: http://www.digipan.net/ See: **PSK31**

digi-peater - A store-and-forward digital repeater station that receives and transmits a data packet on the same frequency. See: **node, packet radio**

digital - Devices that employ calculations done directly with digits instead of measurable physical quantities, such as with analog devices or communication modes. Digital signals operate using alternation between two levels that correspond to either a digit of 1 or zero. See: **analog**

digital modes – Communication that employs digital means of transferring data. In Ham Radio, these include PSK31, JT-65, RTTY, D-STAR, and more. See: **Clover, D-STAR, JT-65, Olivia, PSK31, radioteletype, RTTY**

digital signal processor/processing – Use of digital means while receiving a radio signal to improve the signal-to-noise ratio or to help hear a signal through interference to assure clearer and more legible communication. On transmit, digital signal processing is employed to improve the quality of the station's audio and transmitted signal. Abbreviated as **DSP.**

DIN plug - An electrical connector that was originally standardized by the Deutsches Institut für Normung, the German standards organization.

diode – A solid state semiconductor electronic component that allows current to pass through it in only one direction.

DIP (*CW abbrev.*) – "Dipole antenna." See: **dipole**

dip meter – An instrument used to determine the resonant frequency of an electronic circuit. See: **grid dip meter**

diplexer - A device that provides isolation between two transmit/receive ports such as antenna outputs on two different transceivers or two different antenna connections on the same transceiver. The typical use in Amateur Radio is to couple two transceivers to the same antenna, allowing an operator to receive on one transceiver and transmit on the other, or to use a single antenna with two or more outputs on the same radio. Example: You have a transceiver with an HF antenna output and a VHF antenna output and want to use a single antenna for both frequency ranges.

dipole - A basic antenna usually made of two equal lengths of wire or tubing ("elements"), joined by an insulator in the middle and fed by a two-conductor feedline at the center. Each conductor is attached to one or the other wire element. See: **antenna, element**

direct current - The flow of electricity in which the current only flows in one direction. See: **AC, alternating current, DC**

directional antenna – An antenna specifically designed to receive or emit a signal in a particular direction or directions. See: **omni-directional antenna**

director – 1. An element on a beam or other directional antenna that is located directly in front of the driven element. It "directs" radio-frequency energy in the desired direction. See: **driven element, reflector**
2. An elected official in the ARRL Field Organization, the top representative on the regional level. See: **ARRL, Field Organization**

direct path – A radio signal that travels directly from the transmitting antenna to the receiving antenna without reflecting or refracting off any other object, hill or the ionosphere. See: **multi-path**

discharge – 1. To use up all the energy provided by a battery.
2. To ground a capacitor as a safety measure to make sure any residual electrical charge stored in the component has been dissipated.

discriminator - The stage in an FM receiver in which the modulation information is recovered from the RF signal. See: **demodulation**

dish – A very directional antenna, usually round or parabolic in shape, often used at very high frequencies or for satellite work.

distortion - An undesired change in a waveform or signal.

distress call – A transmission that indicates a life-threatening situation exists. The call usually uses the terms "SOS" on Morse code or digital modes or "Mayday" on voice modes. Distress calls always have priority over any other traffic or activity that might be going on at the time. Issuing a false distress call is illegal. See: **Mayday**, **SOS**

DN, DWN 1. (*CW abbrev.*) – "Down."
2. Sent by a DX station to indicate that he or she is listening down in frequency for calls rather than on the operator's transmit frequency. Example: "DN 2" means the station is listening down approximately 2 kilohertz for stations to call. See: **down**, **DX**, **split**, **UP**

DominoEX – A type of digital mode that uses rapid frequency-shift keying and is nominally twice as fast as PSK31.

dongle (*slang*) – A small piece of hardware that attaches to a computer, television, Ham rig, or other electronic device in order to enable additional functions.

Doppler effect, Doppler shift – The change in frequency of a wave (or other periodic event) to an observer as the source moves relative to the position of the observer. In Amateur Radio, this is most often noted when receiving a signal from an orbiting satellite, requiring the listener to continually vary the receive frequency to keep the satellite tuned in.

double (*slang*) – When two or more stations inadvertently transmit simultaneously on the same frequency or channel.

double bazooka - A type of dipole antenna that uses coaxial cable as its two elements. The shield of the cable is the radiating element and the center conductor acts as a matching transformer to give the antenna a wider bandwidth than a typical dipole. See: **bazooka, coax**

double nickels (*slang*) - A term often used on DX or county-hunting nets or in normal contacts meaning a signal report of "55." Origin: Citizens Band radio. See: **county hunter, RST, signal report**

doublet – Another word for dipole antenna. See: **dipole**

doubling (*slang*) – Two or more stations transmitting simultaneously on the same frequency or channel.

down (*slang*) – An indication by a highly sought DX station that the operator will be listening down in frequency for calls. This keeps the large number of callers from interfering with the DX station. Example: "I'm listening down five." See: **DN, DWN, DX, split, Up**

downlink– The frequency a repeater station or satellite uses to transmit to a user. See: **uplink**

DR (*CW abbrev.*) - "Dear." "My friend." Example: "TNX QSO DR BOB."

Drake – A former manufacturer of Amateur Radio Equipment. The R. L. Drake Company.

D-region, D-layer – In the Earth's atmosphere, the lowest region of the ionosphere, located approximately 25 to 55 miles above the planet. It disappears very quickly after sunset and is slow to regenerate after sunrise, especially on short winter days. The D-layer's primary effect on radio propagation is to absorb energy from the signals as they pass through it.

drift (*slang*) – A slow unintentional and undesired change in the frequency of a transmitter or receiver.

drive – 1. (*noun*) The RF power applied by a transmitter's internal oscillator to the radio's final amplifier stage.
2. (*verb*) To use an internal oscillator to supply RF power to a transmitter's final amplifier stage.
3. (*verb*) To provide RF from a transmitter to an external amplifier to achieve higher power output. See: **driver, oscillator**

driven element – The antenna element that is typically connected directly to the feed line. This term is most often used in beam or other directional antenna systems. See: **beam, director, reflector, Yagi**

driver – 1. The stage in a transmitter that initially provides RF to the radio's internal final power amplifier section.

2. The transmitter/transceiver being used to provide RF to an external power amplifier.

3. An electrical circuit or electronic component that is employed to control another circuit or component.

drop-in charger – A device for restoring energy in a rechargeable battery for a hand-held transceiver. The radio or its detached battery can simply be dropped into a slot in the device until the battery is fully charged. See: **charger**

dropping out (*slang*) - A repeater station is no longer able to hear a station and is unable to attempt to retransmit the signal. Example: "Sorry, you are dropping out of the repeater and I can't copy you."

DSP – Digital signal processor or digital signal processing. See: **digital signal processor/processing**

D-STAR – Abbreviation for Digital smart technologies for Amateur Radio. D-STAR is a digital voice and data protocol specification mostly used in VHF/UHF transceivers manufactured by Icom. As of this writing, other manufacturers hint they may soon produce D-STAR gear. Visit: **www.icomamerica.com/en/products/amateur/dstar/default.aspx** See: **digital, digital modes**

DSW (*CW abbrev.*) – "Goodbye," when conversing with a Russian Ham. Abbreviation for *dos vadanya*, Russian for "Until we meet again."

DTCS – Abbreviation for digital tone-coded squelch. A selective call system in which tones are used to allow the incoming signal to overcome the restrictive squelch of a receiver. See: **squelch**

DTMF – Abbreviation for dual tone multi-frequency, commonly referred to as "touch-tone." Dual audio tones used to send and receive numeric data, such as telephone numbers, or for actuating remote- control commands.

dual band – An antenna or radio that is designed for use on two different Ham Radio bands. See: **multi-band**

dummy load – A non-radiating load connected to a transmitter for testing without interfering with others.

dupe (*slang*) - A duplicate contact. The term is used in a contest in which the rules forbid contacting the same station more than once for score credit. Example: "Sorry but we worked before on this band. You are a dupe." Pronounced "dewp." See: **contest, radiosport**

duplex – An operating mode in which the transmit and receive frequencies are not the same, such as with repeater stations. See: **simplex**

duplexer - A device that allows a single antenna to transmit and receive simultaneously. The device provides isolation on a system on the same band. Example: a repeater receives a signal on 146.28 megahertz at the same time that it is re-transmitting that signal on 146.88 megahertz. A duplexer makes it possible to do this without the transmitter interfering with the receiver.

duty cycle - The proportion of time during which a component, device, or system is operated. Example: The duty cycle of a transceiver is much shorter when using constant-carrier modes such as FM or RTTY as opposed to intermittent modes such as SSB.

DVM - Digital voltmeter.

DWN (*CW abbrev.*) – "Down." See: **down, DN**

DX – 1. (*CW abbrev.*) – "Distance."
2. (*slang*) (*noun*) A distant station or a station outside one's own country or region.
3. (*slang*) (*verb*) To contact stations in other countries via Ham Radio. Example: "I like to DX on 17 meters."

DXCC – DX Century Club. See: **DX Century Club**

DX Century Club - An operating award sponsored by the American Radio Relay League. Abbreviated as DXCC. To earn the award, you must contact at least 100 different countries or DX entities and then confirm those contacts through Logbook of the World or by receiving QSL cards for each. Endorsements are also available for confirming contact with more DX entities. Visit: **http://www.arrl.org/dxcc** See: **DX entities, Logbook of the World, LOTW, QSL card**

DX cluster - A web site on which stations report hearing ("spotting") or making contact with other stations. This allows operators who wish to talk to those stations to go to that frequency and attempt to make a contact. See: **cluster, DX spotting, spotting**

DX Engineering - A vendor of Amateur Radio equipment and supplies. Visit: http://www.dxengineering.com/

DX entity – A location designated as a "country" for the purpose of operating awards or contesting. For example, Alaska, Hawaii and Puerto Rico are all considered to be "countries" or DX entities.

DXer (*slang*) – 1. An Amateur Radio operator who actively pursues contacts with Amateur Radio stations in other countries, and especially in those countries in which there are few operators and with whom contacts are rare.
2. A hobbyist who listens for shortwave broadcasts from stations in other countries and attempts to obtain confirmations from those stations. See: **shortwave listener**

DXLab - A popular software program for logging Amateur Radio on-air contacts and for use in radiosport events. The program is offered free of charge. Visit: http://www.dxlabsuite.com/

DX-pedition, DXpedition – An organized Amateur Radio operation from a foreign country, usually an entity in which there are few native Hams, giving operators an opportunity to establish and confirm contact with that country.

DX spotting - A process in which stations report hearing ("spotting") or making contact with other stations to a web site. This allows operators who wish to talk to those stations to go to that frequency and attempt to make the contact. See: **cluster, DX cluster, spotting**

DX Store, The – A vendor of Amateur Radio equipment and supplies. Visit: **http://www.dxstore.com/**

DX University - A multi-media program offering information, instruction and learning opportunities for DXers. Visit: **http://www.dxuniversity.com/** See: **DX, DXer**

DX window (*slang*) - A range of frequencies set aside by gentlemen's agreement to allow for foreign Amateur Radio stations to call CQ and work other stations around the world while those in the United States and Canada refrain from other types of activity there. The DX stations may also invite U.S. and Canadian stations to call them. Example: 3.790 – 3.800 megahertz is the 75 meter DX window. Visit: **http://www.bandplans.com/**

dynamic range - The ratio between the largest and smallest possible values of a changeable quantity, such as in radio signals or sound reproduction. In Amateur Radio, this most often involves how a receiver handles very weak or very strong signals.

E

Echo

Dit

E – Symbol for electromotive force or voltage, measured in volts.

EE (*cw abbrev.*) (*slang*) – A final salute at the end of a Morse code contact, like a goodbye wave, which is then echoed by the other station. Sent as "Dit dit."

earth ground - A wire conductor that terminates in the earth for electrical purposes, typically using a ground rod. See: **chassis ground, ground, grounding**

Earth-moon-Earth - Using the moon as a passive reflector off which to bounce signals back to Earth at some distance from the originating station. Sometimes referred to as moonbounce. Abbreviated as EME.

earth station - An Amateur Radio station located on or near the Earth's surface that communicates with stations such as the International Space Station, or to talk with other stations on Earth by means of satellites in orbit.

echo – An instance when a received signal arrives at the same location but concurrently by different paths, one taking slightly longer to travel than the other, thus creating what sounds like an echo.

Echolink – A software system that allows licensed Amateur Radio stations to communicate with one another over the Internet through properly equipped repeater stations, using streaming-audio technology and a computer or telephone device. Visit: **http://www.echolink.org/**

eHam.net – A popular web site for Amateur Radio operators. See: **www.eham.net**

EHF - Extremely High Frequency, typically 30 to 300 gigahertz. See: **ELF, HF, UHF, VHF, VLF**

EL (*CW abbrev.*) – "Element." Example: "Using 2 EL beam."

E-layer - The region of the ionosphere in the Earth's atmosphere that is located approximately 55 to 90 miles above the planet. This layer normally absorbs radio signals passing through it. During certain times of the year, however, this layer becomes ionized and will refract signals on higher frequencies, propagating them over greater distances than usual. See: **sporadic-E**

Elecraft – A major manufacturer of Amateur Radio equipment, headquartered in Aptos, California. Visit: **http://www.elecraft.com/**

Electric Radio Magazine – A magazine primarily for Amateur Radio and electronic hobbyists that concentrates on vintage equipment and radio history. Visit: **http://www.ermag.com/**

electrolytic capacitor - A type of capacitor used most often as power supply filters. Such capacitors do require attention to their polarity and are capable of storing enough of an electric charge to be dangerous, even after they have been disconnected from power. Some types of electrolytics also tend to dry out over time and may fail if they have not been used in a long time.

electronic keyer - A circuit used for creating the dots and dashes of Morse Code semi-automatically using a key that is commonly referred to as a "paddle." Dots are made by pressing the paddle one way and dashes by pressing the paddle the other. See: **keyer, paddle**

electromagnetic pulse - A high-energy magnetic field which can be caused by a number of natural occurrences such as a lightning strike or by events such as a nuclear explosion. It could cause severe damage to electronic equipment, power distribution systems, and more. See: **EMP**

electromotive force - The force or pressure that pushes an electrical current through a circuit. Voltage. Abbreviated EMF and represented mathematically by the letter "E." See: **voltage**

element – The conductive parts of an antenna. In a dipole, these are the two wires that are separated by an insulator in the middle. In a beam, these are the parallel conductors attached perpendicular to the boom. See: **beam, boom, dipole**

elephant (*slang*) - a repeater station that receives from a much greater distance than it can transmit, so named because the animal has big ears but a small mouth!

elevation – 1. When describing a beam antenna, the angle at which it is aimed in reference to the horizontal.
2. How high an antenna, tower, satellite or other object is, often referenced in communications to either sea level or height above average terrain in the nearby area. See: **HAAT**

ELF - Extremely low frequency. See: **extremely low frequency**

elmer (*slang*) – An Amateur Radio mentor, usually an experienced operator who assists newcomers to the hobby.

EmComm – Emergency communications. See: **emergency, emergency traffic**

EME - Earth-moon-Earth. See: **Earth-moon-Earth**

emergency – An instance in which there is an imminent threat to life or property.

emergency traffic - A message or communication passed along from one station to another that involves the life or safety of persons. Such traffic always has priority and all stations must standby until it has been completed. See: **informal traffic, NTS, National Traffic System, priority traffic, traffic**

EMF - Electromotive force, another word for electrical voltage. See: **electromotive force, voltage**

EMI - Electromagnetic interference, a disturbance that affects an electrical circuit due to radiation emitted from an external source.

emission – An electromagnetic signal.

emission mode, emission type – The specific type of electromagnetic signal being generated by a transmitting device. Examples: AM, FM, or single-sideband. Emission modes are more completely described and defined by regulatory agencies in each country, such as the Federal Communications Commission in the USA.

EMP - Electromagnetic pulse. See: **electromagnetic pulse**

EPROM, EEPROM – Electrically erasable programmable read-only memory. A digital chip designed for storing and accessing data.

eQSL – A website on which contacts between two Amateur Radio stations may be confirmed electronically and digital QSL "cards" exchanged. Visit: **https://www.eqsl.cc/** See: **QSL card**

ERP - Effective radiated power. The total amount of radio-frequency power being emitted at the antenna considering actual power from the final amplifier stage, feedline loss, and antenna gain.

ES (*CW abbrev.*) – "And."

E-skip – Propagation of signals using sporadic-E-layer refraction. See: **E-layer, sporadic E**

ESSB – Extended single-sideband. See: **extended single-sideband**

ether (*antiquated term*) (*slang*) – Prior to the discovery of the ionosphere's effect on radio waves, theory held that signals were propagated by a substance called ether. The term is still used as jargon for signal propagation.

EU (*CW abbrev.*) – "Europe."

EVE (*CW abbrev.*) – "Evening."

exam cram (*slang*) – A one- or two-day session in which instructors go over and explain correct answers to Amateur Radio exam pool questions with prospective Hams. At the end of the session(s), Volunteer Examiners administer the licensing exams. See: **question pool, Volunteer Examiner**

exam question pool – See: **question pool**

exam session - An event in which the Amateur Radio license examination is administered by Volunteer Examiners. See: **exam cram, VE, Volunteer Examiner**

exchange – Information passed between two Ham stations while in contact with each other. Might include signal reports, name, location, and equipment being used plus more. A far shorter exchange is typically used in radiosport. Example: "The contest exchange consists of signal report and state." See: **radiosport**

exciter – 1. The oscillator and—if using voice modes—the modulator in a large transmitter.

2. A transmitter or transceiver used to drive an external power amplifier.

extended single-sideband - Experimentation with the audio quality of single-sideband transmissions to attempt to achieve high fidelity sound within the limitations of the mode. Abbreviated as ESSB. Visit: http://www.nu9n.com/essb.html

Extra class (*slang*) – The Amateur Extra class of Amateur Radio license. See: **Amateur Extra**

Extremely Low Frequency – Part of the radio-frequency spectrum, usually designated as 3 to 30 hertz. While ELF has use in weather science and the medical field, it is generally thought of in relation to communication with submarines. See: **ELF**

eyeball, eyeball QSO (*slang*) - A face-to-face meeting between two Ham Radio operators.

EZNEC – Brand name of a popular software program for modeling antenna design. Visit: http://www.eznec.com/ See: **antenna modeling, Numerical Electromagnetics Code**

F

Foxtrot

Di – di – dah – dit

FAA - Federal Aviation Administration. See: **Federal Aviation Administration**

fading – A reduction in the strength of a received signal due to atmospherics or other factors.

false signal – Any transmission by an operator, whether properly licensed or not, that is intended to mislead or confuse. Example: Someone reporting an emergency when none exists. Such transmissions are illegal. See: **deceptive signal**

Family Radio Service – An unlicensed communications service for short-range communications by friends or family members using hand-held radios. The service uses the UHF range so does not suffer the interference that plagues the 11-meter Citizens Band. Abbreviated as FRS.

fan dipole – An antenna that uses one feed point for several dipole antennas, each cut to operate on a specific frequency band. Each shorter antenna fans out beneath the one above it, thus the name.

Farnsworth method - a way of learning and sending Morse code characters in which each character is sent at a faster rate but spaces are left between each character to effectively lower the word-per-minute rate. Example: Characters are sent at 15 words per minute but the spacing is adjusted so that the overall code speed is 5 words per minute. This often helps those learning Morse code to increase reception speed more quickly as they practice.

fast-scan television – A mode that allows Amateur Radio operators to send and receive live-action TV images similar to analog broadcast television. See: **ATV, FSTV**

FB (*CW abbrev.*) – "Fine business," "Excellent," "Good," "Fine."

F/B – Front-to-back. See: **front-to-back ratio**

F connector – An inexpensive coax connector, used primarily on smaller size cables for television and cable TV.

FCC - Federal Communications Commission. See: **Federal Communications Commission**

FDIM – Four Days in May. See: **Four Days in May**

Federal Aviation Administration - The agency in the USA that regulates all facets of air transportation. This includes such areas as tower height restrictions and lighting requirements. Abbreviated as FAA.

Federal Communications Commission - The governmental agency that regulates Amateur Radio in the USA. Abbreviated as FCC.

Federal Registration Number - An identification number assigned to individuals by the Federal Communications Commission for use when accessing the Commission's on-line site and user accounts. Abbreviated as FRN. The number can be used while applying for, modifying or renewing an Amateur Radio license.

feeder - A transmission line used to transfer the power from a transmitter to the antenna. Usually a coaxial cable or open-wire line. Often referred to as a feedine. See: **feedline, transmission line**

feedline – The wire or cable that connects a radio to an antenna.

fender mount – A device for affixing an antenna to the fender of a vehicle.

FET - Field-effect transistor. A type of transistor that is typically used as an amplifying device.

FER (*CW abbrev.*) – "For."

Field Day – An annual Amateur Radio operating event in June in which groups set up stations in portable locations to practice emergency communications. Field Day is sponsored by the American Radio Relay League and is the hobby's largest event in terms of participants. See: **operating event**

Field Offices – Branch offices of the Federal Communications Commission, which is located in Washington, DC. Regional offices are in Chicago, Kansas City, and San Francisco. There are 16 district offices and 8 resident agent offices around the country. Personnel in these offices conduct on-scene investigations and inspections and audits, respond to safety of life issues, and investigate and resolve interference complaints and violations in all communications services. See: **Federal Communications Commission**

Field Organization – Volunteers who work with the American Radio Relay League to assist other Hams. Some are elected by ARRL members while others are appointed and represent constituents at the regional, state/section and local levels. See: **ARRL**

field strength meter - A test instrument used to show the presence of and measure the strength of radio frequency energy in the area where the meter is located.

filter – A circuit or device designed to only allow certain specific frequencies to pass through.

final – 1. The last transmission by a station during a contact. Example: "This will be my final. Good night!"
2. The last amplifying stage of a radio transmitter or external amplifier.
3. The amplification tube or transistor that is a part of a transmitter's final output stage. Example: "The finals in my amp are a pair of 811As."

first personal (*slang*) – An operator's first name. This term originated as Citizens Band jargon but is usually frowned upon by Hams. Most prefer simply "name" or "handle."

fist (*slang*) – The personal and individual characteristics of how an operator sends Morse code characters. An operator with a good "fist" is one whose sending is easily de-coded.

FISTS - An organization of CW enthusiasts whose mission is to promote the use of Morse code on the Amateur Radio bands, encourage newcomers to learn and use CW, and to sponsor activities to increase interest in the mode. Visit: **http://www.fistsna.org** See: **CW**

five-by-five (*slang*) – A signal report giving a signal strength of 5 and a readability of 5 on the RS scale. Sometimes referred to as "double nickels." See: **double nickels, RST**

fixed station - A radio station that is designed to be operated from a fixed location instead of being portable or mobile (in a vehicle). Sometimes called a "base" or "base station." See: **base station, mobile, portable**

flat-top antenna (*antiquated term*) (*slang*) – A dipole antenna hung so the ends and middle are at the same height above the ground. See: **dipole**

flat topping – Over-modulating a carrier signal so badly that it distorts the waveform when viewed on an oscilloscope.

F-layer- The region of the ionosphere between about 90 and 400 miles above Earth. This area of the Earth's atmosphere is responsible for most long distance radio signal propagation on the frequency bands below 30 megahertz. During hours when the Earth is in sunlight, solar heating can cause the F-layer to split into two separate layers, F1 and F2.

FLdigi – Software digital modem for receiving and sending Amateur Radio digital modes. Visit: **http://www.w1hkj.com/** See: **digital modes**

flea market – An area at a hamfest or other gathering where used Amateur Radio equipment, parts, and other items may be bought and sold. See: **boneyard, hamfest**

Flexradio Systems – A major manufacturer of software defined radios for Amateur Radio, headquartered in Austin, Texas. Visit: **http://www.flexradio.com/**

flutter - Rapid variation in the strength of a received signal, normally caused by a variation in the propagation of the signal.

FM – 1. Frequency modulation. See: **frequency modulation**
2. (*CW abbrev.*) – "From."

FMRE - Mexican Federation of Radioexperimentadores. See: Mexican Federation of Radioexperimentadores

FOC - First Class CW Operators Club. A worldwide group of Amateur Radio operators that attempts to promote good CW (Morse code) operating habits, activity, friendship and socializing on the bands. Visit: **http://www.g4foc.org**

folded dipole – A dipole antenna in which the ends of the wires have been folded back and run parallel to the original wires, forming a very "skinny" loop. Television twin-lead or ladder line is often used in construction. The antenna typically offers wider bandwidth than a traditional dipole. See: **dipole, twin-lead**

formal traffic (*slang*)– A message or communication that follows established format and is passed along from one station to another in a pre-prescribed manner. Such traffic may also carry a "precedent," such as "routine," "priority," or "emergency. See: **emergency traffic, informal traffic, NTS, National Traffic System, precedence, priority traffic, traffic**

forward power – The power radiated by an antenna less any reflected power. See: **antenna analyzer, reflected power, SWR, standing wave ratio**

Four Days in May – An annual seminar specializing in QRP interests, sponsored by the QRP Amateur Radio Club International. It is held in Dayton, Ohio, and coincides with the Dayton Hamvention. Abbreviated as FDIM. Visit: **http://www.qrparci.org/fdim** See: **Dayton, QRP ARCI**

fox hunt – Using radio direction-finding equipment to find a hidden transmitter. Such activities are usually conducted as a fun exercise or contest but skills learned can come in handy if an illegal, stolen, or hung-up transmitter is detected and needs to be located. See: **bunny hunt, RDF**

FREQ (*CW abbrev.*) – "Frequency."

frequency – Generally, the rate of oscillation of a wave. The frequency of audio and radio waves is measured in hertz (cycles per second). See: **hertz, oscillator**

frequency coordinator – A group or an individual responsible for coordinating and assigning frequencies for repeater stations, assuring minimum interference to existing repeaters.

frequency modulation - A mode in which the encoding of information on a carrier wave is accomplished by varying the instantaneous frequency of the wave. Abbreviated as FM.

frequency shift keying - A methodology in which digital information is transmitted through discrete frequency changes of a carrier wave. Abbreviated as FSK. In Amateur Radio, FSK has primarily been used with such digital modes as PSK31 or RTTY but is being replaced more now with audio frequency shift keying or AFSK. See: **AFSK, audio frequency shift keying, FSK, PSK31, radioteletype, RTTY**

Friendly Candy Company (*slang*) – Term sometimes used for the FCC, the Federal Communications Commission. See: **FCC**, **Federal Communications Commission**

FRN – Federal Registration Number. See: **Federal Registration Number**

front end – The circuitry in a receiver that first encounters a signal at the input, usually including all components that act on an incoming signal before its frequency is altered for more processing.

front-end overload – When a very strong signal overpowers a receiver's early stages, causing interference.

front end protector – A circuit which can guard against a receiver being overloaded or damaged by very strong signals appearing at its input. Of special value in a situation when other Amateur Radio stations are operating very close by.

front-to-back ratio - The ratio of an antenna's gain in the forward direction to that in the opposite direction. Most often a factor in beam antennas and directional arrays.

front-to-side ratio - The ratio of an antenna's gain in the forward direction to the gain at right angles—off the sides—to the forward direction. Most often a factor in beam antennas and directional arrays. Abbreviated as F/S.

FRS – Family Radio Service. See: **Family Radio Service**

F/S – Front-to-side. See: **front-to-side ratio**

FSK - Frequency shift keying. See: **frequency shift keying**

FSTV - Fast scan TV. See: **ATV, fast scan TV**

full break-in – Employing circuitry when using Morse code (CW) to be able to receive signals, even between dots and dashes, while transmitting. This allows the receiving station to interrupt the communication without waiting for the transmitting station to finish sending. This is sometimes referred to as "QSK," from the Q signal for "I can operate break-in." See: **break-in, QSK, Q signal, semi-break-in**

full duplex – A mode of operation in which communication takes place on two different frequencies simultaneously. An example is a typical telephone conversation.

full gallon (*slang*) – Running the maximum amount of legal power from an amplifier.

full quieting – A condition on FM transmissions in which an incoming signal is sufficient to completely cover any other noise in the receiver.

FUNcube – A series of educational Amateur Radio satellites built and maintained by Hams in Great Britain, the Netherlands and others under the banner of AMSAT-NA and AMSAT-UK. Visit: **http://funcube.org.uk/** See: **AMSAT, CubeSat**

fundamental - The lowest frequency or band of frequencies to which a harmonic frequency or band is related. See: **harmonic**

fuse – A safeguard component designed to fail and open the circuit if the current being drawn in an electrical circuit exceeds the maximum rated current for the device. The fuse must be replaced with a new one once the cause of the excessive current is corrected. See: **circuit breaker**

G

Golf

Dah – dah – dit

G (*slang*) – An Amateur Radio operator from Great Britain. Many British call signs begin with the letter "G." Example: "I had a nice chat on 40 meters with a 'G' from Liverpool."

G5RV – A form of dipole antenna that uses a specific length of balanced feedline between the feedpoint and the rest of the coax feedline as a matching stub in order to achieve a better match on several bands. Originally developed by British Amateur Louis Varney G5RV. See: **balanced line, match, matching stub**

GA - 1. (*CW abbrev.*) – "Go ahead."
2. (*CW abbrev.*) "Good afternoon."

gain – 1. In antennas, the increase in the effective power radiated by or in received signal strength from an antenna in a certain desired direction or angle. See: **beam**
2. Increase in received or transmitted signal strength from an amplifier.

gallon (*slang*) – Transmitter output power. Example: "I'm running a half gallon here." (750 watts, or half the maximum allowed, 1500 watts PEP, a full gallon.) See: **full gallon**

GB (*CW abbrev.*) – "Goodbye."

GD (*CW abbrev.*) – 1. "Good day."
2. (*CW abbrev.*) "Good."

GE (*CW abbrev.*) – "Good evening."

gel cell - A sealed lead-acid rechargeable battery which uses a chemical gel to generate an electrical charge.

General class – The current intermediate class of Amateur Radio license available in the USA. See: **Amateur Extra, Technician**

general-coverage receiver - A receiver that is capable of hearing a very wide range of frequencies. Typically, a general-coverage receiver will tune continuously from the AM broadcast band (535 kilohertz) to 30 megahertz, including the Ham bands. See: **Ham-band-only receiver**

General Mobile Radio Service - A licensed North American FM UHF radio service for short-distance two-way communication for family use, typically within a city. Abbreviated as GMRS.

generator - A device that converts mechanical energy to electrical energy for use in an external circuit. In Amateur Radio, this is typically a gasoline-powered internal combustion engine that produces 110 volts AC current for use in situations when normal commercial power is unavailable.

GESS (*CW abbrev.*) – "Guess."

Gigaparts – A vendor of Amateur Radio equipment and supplies. Visit: **http://www.gigaparts.com/**

gin pole – A device attached temporarily to a tower to enable raising and lowering tower sections while constructing the tower or antennas or other objects while attaching or detaching them from the structure. It usually includes a rope or wire long enough to reach the ground below, a winch or pulley, and a strong clamp for attaching the device to the tower while in use.

GG (*CW abbrev.*) – "Going."

GL (*CW abbrev.*) – "Good luck."

GM (*CW abbrev.*) – "Good morning."

GMRS – General Mobile Radio Service. See: **General Mobile Radio Service**

GMT – Greenwich Mean Time. See: **Greenwich Mean Time**, **Coordinated Universal Time**, **Zulu time**

GN (*CW abbrev.*) – "Good night."

GND (*CW abbrev.*) – "Ground."

go ahead (*slang*) – "I am completing my transmission. Go ahead with yours."

go-kit (*slang*) - A complete self-contained portable station including transceiver, power supply and/or battery, antenna and personal supplies that allows a Ham to quickly go and set up for a portable operation or to report where needed in an emergency situation.

Gooney Bird (*slang*) (*antiquated term*) – A small, low-power two-meter AM transceiver once manufactured by the Gonset Company.

GOTA station – "Get on the Air" station. A station set up for ARRL Field Day to be operated by unlicensed visitors and newcomers to the hobby under the supervision of a control operator. See: **Field Day**

grace period (*slang*) - The time period allowed by the Federal Communications Commission after an Amateur Radio license has expired in which the licensee may still renew that license. No on-air operation is allowed during the grace period. If the license is not renewed by the end of the period, the examination must be retaken to once again become licensed.

gray line – The transition region between day and night or night and day that constantly moves around the globe. Signals are often strongest when both ends of a conversation are in this day/night or night/day zone.

great circle route - The shortest path between any two points on the planet Earth. This would be the direction an operator would typically want to aim a beam antenna to send and receive the strongest signal to that area. See: **long path**

green stamp (*slang*) – One dollar bill in USA currency. Generally one or two dollars along with a self-addressed envelope are sent along with a QSL card request to an operator in a foreign country in order to offset postage for a return card. (Note: the self-addressed envelope accompanying the request is not stamped since USA-issued stamps are not honored in most countries.) See: **IRC**

Greenwich Mean Time - Coordinated Universal Time. The time at 0-degrees longitude, which passes through Greenwich, England. This time is generally used by Amateur Radio operators when logging contacts in order to avoid confusion brought on by differences in time zones around the world. See: **UTC, GMT, Greenwich Mean Time, Zulu time**

grid dip meter - An instrument used to determine the resonant frequency of an electronic circuit. See: **dip meter**

grid square - An alphanumeric geographical coordinate system (usually given as four or six characters), based on the Maidenhead Locator System developed by VHF/UHF enthusiasts. The entire globe is divided into rectangles denoted by alphanumeric codes. These grid squares are often used in VHF and UHF DX activities and many radiosport events. Visit: **http://www.arrl.org/grid-squares**

ground - 1. An electronically neutral circuit that has the same electrical potential as the earth that surrounds it. See: **earth ground**
2. A non-current carrying circuit designed to provide electrical safety.

3. A point of reference within an electrical circuit or system. See: **chassis ground**
4. The negative side of an electrical circuit.
5. The negative side of a battery.

grounding – electrically connecting equipment or antennas to an earth ground. See: **earth ground**

ground-plane antenna - a vertical antenna typically built with the vertical radiating element one-quarter-wavelength long and with several radials slightly longer than one-quarter wave extending horizontally from the base.

ground rod - A highly-conductive rod that is driven into the earth to create a ground for electrical equipment. For Amateur Radio stations, a heavy copper wire or strap is run from the station equipment—which is bonded together with strap—to the ground rod. See: **bond**

ground wave - radio waves that tend to travel along the surface of the earth, often beyond the horizon.

GUD (*CW abbrev.*) – "Good."

guy – A set of ropes or wires used to secure a tower or mast once it has been erected. Sometimes mispronounced as "guide wires."

GV (*CW abbrev.*) – "Give."

H

Hotel

Di – di – di – dit

HAAT – Height above average terrain. A designation of elevation often used with antennas installed on very tall towers.

half-duplex - A communications mode in which a radio transmits and receives on two different frequencies but performs only one of these operations at a time. An operator using a half-duplex system would transmit and then listen for the other station's transmission. See: **duplex, full duplex, HDX**

half-wave antenna - An antenna consisting of a length of wire or conductive metal tubing that is electrically one-half wavelength long for the desired operating frequency. See: **resonance, wavelength**

Hallicrafters – A former manufacturer of Amateur Radio equipment.

ham (*slang*) – An Amateur Radio operator. Though the origin of the term is not certain, it is believed telegraphers in the early days of the use of Morse code called newcomers "hams."

ham-band-only receiver--A receiver which will only tune the bands assigned to or used by Ham radio operators. See: **general-coverage receiver**

ham bands – The frequencies on which Amateur Radio operations are authorized by the various agencies in countries around the world.

Ham City – A vendor of Amateur Radio equipment and supplies. Visit: **http://www.hamcity.com/**

hamfest – A gathering of Amateur Radio enthusiasts at which Hams meet to buy, sell, and swap equipment, visit with each other, and attend seminars on various subjects of interest to hobbyists. See: **boneyard, flea market**

Hammarlund – A former manufacturer of Amateur Radio equipment, best known for its line of receivers.

Ham Nation – A weekly television show dealing with Amateur Radio, streamed live via the Internet. Past shows are archived as well and available for viewing. Visit: **https://twit.tv/shows/ham-nation**

ham operator – A person who holds a valid Amateur Radio operator's license from his or her country's communications governing agency.

Ham Pros – A group of independent vendors of Amateur Radio equipment and supplies spread around the country including Associated Radio, Lentini Communications, Radio City, and Universal Radio. Visit: **http://www.hampros.com/**

Ham Radio Deluxe - A commercially available software system for Amateur Radio Operators which offers computer control of most transceivers, an electronic logbook, and ability to operate most digital modes. Visit: **http://www.ham-radio-deluxe.com/**

Ham Radio Magazine (*antiquated term*) – Amateur Radio magazine published from 1968 until 1990. Known for its emphasis on more technically-oriented content.

Ham Radio Outlet – A multi-location vendor that sells Amateur Radio equipment and supplies. Abbreviated as HRO. Visit: **http://www.hamradio.com/**

ham shack (*slang*) – The area in which a Ham has set up his radio station. It can be a separate building but is usually just a corner in a room, garage, or basement area.

Ham Station, The – A vendor of Amateur Radio equipment and supplies. Visit: **http://www.hamstation.com/**

hand-held (*slang*) - A small, battery-powered transceiver, usually for the VHF and/or UHF frequencies, so named because it is small enough to be carried easily in one hand. Sometimes called a brick, walkie-talkie or HT.

HandiHam System – An organization for Amateur Radio operators with physical disabilities. Visit: **http://www.handiham.org/**

handi-talkie (*slang*) - A small, hand-held battery-powered transceiver, usually for VHF or UHF frequencies. Sometimes called a brick, walkie-talkie or HT.

handle (*slang*) - A Ham Radio operator's name. Most now prefer that "name" be used instead of handle or the Citizens Band terms "personal" or "first personal." On CW, the abbreviation "OP" is typically used instead of "name" or "handle."

hang time (*slang*) – The length of time following the end of a transmission on a repeater station before the repeater carrier drops. This short pause allows others who want to access the repeater a chance to do so before the signal drops or another station transmits. On many repeaters, a courtesy beep will alert users when the repeater is ready to accept another transmission.

harmonic – 1. The multiple of a fundamental frequency.

2. An undesirable spurious emission that occurs on a multiple of the frequency of the original transmission. This is a violation of rules and is even more serious if the harmonic transmission falls outside an Amateur Radio band, possibly causing interference to other services. See: **spurious emission**

2. (*slang*) The children of an Amateur Radio operator.

HDX - Half-duplex. See: **half duplex**

headphones – A small speaker or two small speakers mounted on a band that can be worn on the head with the speaker(s) on each ear. The speaker lead(s) is then plugged into the headphones outlet on the receiver/transceiver. See: **headset**

headset – Headphones with the addition of a small microphone on a boom positioned near the lips. Sometimes called boom-set. See: **boom-set, headphones**

health and welfare traffic – Messages originating from a disaster area dealing with the well-being of those in the affected area, intended for family or friends who might be concerned about them. Health and welfare messages have lower precedence than emergency or priority traffic and must wait until that traffic has been completed. See: **emergency traffic, precedence, priority traffic**

Heathkit – At one time, a leading manufacturer of Ham Radio equipment, with most products offered as kits.

Heil – A manufacturer of equipment for Hams, best known for its line of microphones and headsets/headphones. Founded by Bob Heil K9EID, a member of the Rock & Roll Hall of Fame for his work with sound techniques in concert venues. Visit: **http://www.heilsound.com/amateur/**

hellschreiber - A digital mode for sending and receiving text on the Amateur Radio HF bands. Often shortened to "hell."

henry – Unit of electrical inductance. See: **inductance**

hertz – The unit for measuring frequency. One hertz is one cycle per second of a repetitive wave, and typically refers to a sound or electromagnetic wave. One kilohertz (kHz) is 1,000 cycles per second. One megahertz (mHz) is one million cycles per second. One gigahertz is one billion cycles per second. Abbreviated as Hz.

heterodyne – 1. (*verb*) Putting two signals of differing frequencies together—intentionally or not—to create a third signal that is an arithmetical factor of the first two. Example: A signal of 3.700 megahertz mixed with a signal of 3.701 megahertz creates and audio signal of one kilohertz. This may be done inside a receiver to change the frequency of a signal to another frequency that may be more easily processed, or to create a tone so CW or SSB may be copied. See: **BFO, integral frequency**

2. (*noun*) A signal created by the mixing of two other signals of different frequencies. Such a signal can be the cause of interference.

hex beam – A popular Amateur Radio two-element wire beam antenna. The antenna uses wire for its elements and supporting spreaders are made of Fiberglas, bamboo or other material to form a two-element directive antenna. This method allows for the beam to have elements covering six through twenty meters all fed with a single coax feedline. Visit: **http://www.karinya.net/g3txq/hexbeam/**

HF – High frequency. See: **high frequency**

HH (*CW abbrev.*) – Error while sending, sent as one character, di-di-di-di-di-di-di-dit.

HI (*CW abbrev.*) – 1. Indicates the operator is laughing. Sometimes used on phone modes as well but generally frowned upon. The operator can simply laugh rather than saying, "Hi!" to indicate amusement.

2. (*CW abbrev.*) – "High."

hi fi audio – See: **extended single-sideband**

high frequency - Typically defined as that portion of the radio-frequency spectrum that falls between 3 and 30 megahertz. This range is also sometimes referred to as "shortwave." Abbreviated as HF. See: **EHF, shortwave, UHF, VHF, VLF**

high-pass filter - A filter designed to pass high frequency signals while blocking lower frequency signals. See: **filter**

holiday style (*slang*) – A Ham operating from a foreign country while on vacation or for work purposes and getting on the air casually as he or she has time or opportunity. Differs from a DXpedition in which operators attempt to be on the air around the clock. See: **DXpedition**

hollow state (*slang*) – Term for equipment that uses vacuum tubes, as opposed to solid state devices.

homebrew (*slang*) – 1. (*noun*) Radio equipment constructed by an individual, not purchased already built.
2. (*verb*) To construct a piece of equipment rather than purchase it commercially built.

home QTH (*slang*) – An operator's city or place of residence. Considered unnecessarily wordy and not encouraged. Comes from the Q-signal for, "What is your location?" Example: "The home QTH here is Springfield, Illinois," or, "I have arrived at the home QTH so I'll sign off now." See: **Q-signals, QTH**

hop (*slang*) – The distance between two stations communicating by reflecting the radio waves off of the ionosphere or some other object or entity. See: **multi-hop**

horizontal loop – A loop antenna that is parallel to the ground. Sometimes called a skywire loop. See: **loop, skywire**

horizontal polarization – 1. An electromagnetic wave which has its electrical lines of force parallel to the ground.
2. An antenna that has its element or elements roughly parallel to the ground beneath it. See: **vertical polarization**

HPE (*CW abbrev.*) – "Hope."

HQ 1. (*CW abbrev.*) – "Headquarters."
2. Abbreviation for American Radio Relay League headquarters in Newington, Connecticut. See: **American Radio Relay League**

HR (*CW abbrev.*) – "Here," "Hear."

HRD 1. (*CW abbrev.*) – "Heard."
2. Ham Radio Deluxe. See: **Ham Radio Deluxe**

HT – Handi-talkie. See: **hand-held, handi-talkie**

Hurricane Watch Net – An Amateur Radio network that goes into continuous session when a hurricane threatens landfall anywhere in North and Central America. Abbreviated as HWN. Visit: **http://www.hwn.org/**

HV 1. (*CW abbrev.*) – "Have."
2. Abbreviation for "high voltage."

HW (*CW abbrev.*) – "How."

HW?, HW NW? (*CW abbrev.*) (*slang*) - "How do you copy my signal now?" Or, "Back to you to see if you still copy my signal."

hybrid rig (*slang*) – A receiver or transceiver that uses both tubes and transistors in its various circuits.

Hz – Abbreviation for the unit of frequency, hertz. See: **hertz**

I

India

Di – dit

I - Symbol for electrical current in a circuit, measured in amperes.

iambic keying - A method of sending Morse code in which the operator closes both keyer paddle connections at the same time to send alternating dots and dashes. An electronic keyer with the proper logic circuitry and an iambic paddle are necessary to do this.

IARU - International Amateur Radio Union. See: **International Amateur Radio Union**

IC - Integrated circuit. A single device that contains several or many other types of electronic circuits inside it. See: **integrated circuit**

Icom - A major manufacturer of Amateur Radio equipment, headquartered in Japan. Visit: **http://www.icomamerica.com/**

ID – (*verb*) To identify a station that is transmitting, typically by giving the assigned call letters. See: **call letters, identify**
(*noun*) A station's call letters as assigned by the pertinent regulatory agency. See: **call letters, call sign**

identify – Giving the assigned call letters of the station at the beginning and end of a transmission or series of transmissions or at least every ten minutes during an ongoing conversation.

IF - Intermediate frequency. See: **intermediate frequency**

IMD – Intermodulation distortion. Spurious emissions that can occur when two or more signals of different frequencies are mixed together in a receiver or transmitter creating additional signals that can create minor to severe interference within the receiver or transmitter.

impedance – Resistance to the flow of electric current, measured in ohms and represented by the letter Z.

inductance - a measure of the ability of a coil to store energy in the form of a magnetic field. Unit of measurement is the henry.

inductor - an electrical component usually composed of a coil of wire wound on a central core. An inductor stores energy in a magnetic field. See: **coil, inductance**

informal traffic – A message or communication passed along from one station to another in a conversational manner. See: **formal traffic, net, precedence**

input, input frequency - The frequency on which a repeater station's receiver listens for signals. This should be the transmit frequency for the operator's transmitter if he or she wishes to communicate through that repeater.

insulator - A material that effectively resists the flow of electric current.

integrated circuit - A single component that contains multiple devices. Abbreviated as IC.

intermediate frequency - A frequency to which a received signal is shifted from its original frequency as an intermediate step in processing that signal. Converting incoming signals to an intermediate frequency can enhance amplification, filtering and signal processing. Abbreviated as IF. See: **heterodyne**

intermod – Intermodulation. When spurious signals are produced by two or more signals inadvertently mixing inside a receiver, such as at a repeater station where multiple transmitters may be located nearby. This can make it difficult for the station receiver to process incoming signals.

intermodulation distortion - Spurious emissions that can occur when two or more signals of different frequencies are mixed together in a receiver creating additional signals that can create minor to severe interference within the receiver.

internal tuner – An antenna matching device that is built into a transceiver. Such devices automatically attempt to find the best match to the antenna system when they are engaged, usually by a button push. Most commercially available Amateur Radio transceivers have an internal "tuner" or offer one as an option. See: **antenna tuner, auto-tuner**

International Amateur Radio Union - A worldwide Amateur Radio organization whose members consist of the official radio societies from all participating countries. In the USA, the ARRL is the country's member organization. Abbreviated as IARU. Visit: **http://www.iaru.org/**

International Morse code – The most accepted version of a method of transmitting text information as a series of dots and dashes sent and received as on-off tones, lights, or clicks that can be directly understood by a skilled listener or observer. The length of a dot is one unit and a dash is three units. Except with the Farnsworth method, the space between each part of a character is one unit, between each letter/number is three units, and between words is seven units. Today, and in Amateur Radio, the International Morse code is the standard version in use. Morse code was named for its inventor, Samuel F.B. Morse. In Amateur Radio, the mode is also often referred to as CW. See: **CW, code**

(Note: The header for each letter in this dictionary contains the "sound" of the appropriate letter in International Morse code, not dots and dashes. It is far better for someone learning the code to hear the way letters sound instead of learning the characters as dots and dashes. The brain has to make an additional translation from visual to sound from dots/dashes.)

International Telecommunications Union - The international body charged with specifying by treaty the worldwide guidelines concerning the use of the electromagnetic spectrum for communications purposes. Visit: **http://www.itu.int/**

Internet Radio Linking Project –A system that links individual amateur radio stations and repeater stations via the Internet using voice-over-IP (VoIP) software and hardware. Abbreviated as IRLP. Visit: http://www.irlp.net/

inverted V antenna - A wire dipole antenna in which the center is supported at a point that is higher than the two ends, forming an upside-down vee. Useful because it takes less space than a typical dipole. See: **dipole**

inverter - An electrical device that converts direct current (DC) to alternating current (AC).

I/O - Input/output.

ionosphere - The electrically charged region of the Earth's atmosphere that can refract radio signals. This region is located approximately 40 to 400 miles above the Earth's surface.

IOTA - Islands on the Air. See: **Islands on the Air**

IRC - International Reply Coupon. A coupon that can be purchased at post offices throughout the world that can be exchanged in foreign countries for return postage for a surface mail letter to the country that issued the coupon. Use and recognition of IRCs are rapidly diminishing.

IRLP – Internet Radio Linking Project. See: **Internet Radio Linking Project**

Islands on the Air - A movement to encourage Amateur Radio operation from and with islands throughout the world. Abbreviated as IOTA. Visit: **http://www.rsgbiota.org**

isotropic – A theoretical single point source of radiated radio frequency energy used to calculate and compare gain of various antennas.

ITU - International Telecommunications Union. See: **International Telecommunications Union**

J

Juliet ("Jew – lee – ETT")

Di – dah – dah – dah

J-38 (*antiquated term*) – Now an antique, a type of simple but sturdy Morse code straight key originally manufactured for railroad telegraph and military use. See: **key, straight key**

jack – A female electrical connector designed to accept a plug. See: **plug**

jam (*slang*) – To deliberately cause intentional interference by transmitting on a frequency already in use by other stations. Such operation is illegal in the Amateur Radio service and can lead to monetary fines and jail time.

Jamboree on the Air - An annual on-the-air event in which Boy Scouts worldwide attempt to make contact with each other using Amateur Radio. Abbreviated as JOTA.

Johnson, E. F. – A former manufacturer of Amateur Radio equipment, best known for their series of Viking transmitters.

JOTA - Jamboree on the Air. See: **Jamboree on the Air**

J-pole – a vertical antenna consisting of a half-wavelength radiator fed by a quarter-wave matching stub. The physical shape resembles the letter "J," thus the name.

JT-65 – One of the digital operating modes available using WSJT software. JT-65 is especially useful when employed for weak-signal radio communication between Amateur Radio operators. The mode is named after Joe Taylor, K1JT, the Nobel laureate who developed the mode. See: **digital**, **digital modes**, **WSJT**

jug (*slang*) – Very large transmitting tubes.

juice (*antiquated term*) (*slang*) – Early term for electrical current.

jumper – 1. A small piece of wire used to electrically connect two different points in a circuit.
2. A short piece of coax cable used to connect transmitters, receivers, and/or transceivers to other station accessories such as meters, tuners, or antenna switches.

junkbox (*slang*) - A collection of spare parts and miscellaneous items retained—not necessarily in a single container—by Hams, makers, or other electronic hobbyists.

jury rig (*slang*) - Fix a problem in a sloppy or unorthodox way.

K

Kilo ("Kee – low")

Dah – di – dah

K (*CW abbrev.*) – "I am finished with my transmission. Go ahead."

K9YA Telegraph – An e-zine for the amateur radio community, offered free on-line in a PDF form. Visit: **http://www.k9ya.org/**

KC (*antiquated term*) – Abbreviation for 1000 cycles per second or kilocycles. Replaced by kilohertz. See: **hertz**

Kenwood - A major manufacturer of Amateur Radio equipment, headquartered in Japan. Visit: **http://www.kenwoodusa.com/**

kerchunker (*slang*) – A person who transmits to activate a repeater station but does not identify his or her station. Such practice is generally discouraged and is technically illegal since the operator does not identify. See: **kerchunking**

kerchunking (*slang*) - Activating a repeater station by making very short transmissions without identifying. Stations sometimes do this to determine if they have a strong enough signal to work through the repeater or not. Those who do are called kerchunkers.

key – 1. (*noun*) A switch or button used to create dots and dashes (dahs and dits) in Morse code.

2. (*verb*) To use such a device to create the dots and dashes and form characters.

3. (*verb*) To engage the push-to-talk switch on a microphone to begin transmitting. See: **key-up, un-key**

key clicks - Undesired "clicks" or "thumps" generated by a CW transmitter as the Morse code key is used to turn the carrier on and off. Clicks can cause interference to other stations operating on the band. See: **clicks**

keyer – An electronic device for creating the dots and dashes of Morse code semi-automatically using a key that is commonly referred to as a "paddle." Dots are made by pressing the paddle one way and dashes by pressing the paddle the other. See: **electronic keyer, key, paddle**

key-up - Activating a repeater station by transmitting on its input frequency. Example: "I was able to key up the 88 repeater from my office downtown."

Kids Day – A twice-a-year operating event in which young Hams and those interested in the hobby are urged to operate a station and talk with each other as well as with other Amateurs. Fathers and mothers are also encouraged to allow their own children to operate. Sponsored by the American Radio Relay League. Visit: **http://www.arrl.org/kids-day**

kilo - The metric prefix meaning multiply the suffix value by 1000.

kilocycle (*antiquated term*) – A frequency of one thousand cycles per second. This term has been replaced by kilohertz (kHz). See: **hertz**

kilohertz – A frequency of one thousand hertz. See: **hertz**

kilowatts – One thousand watts. See: **watt**

K-index - A measure of the Earth's magnetic field. Lower numbers typically mean better propagation of radio waves.

KN (*CW abbrev.*) - "I am finished with my transmission. The station to which I am talking may go ahead but no breakers, please." Sent as a single character, dah-di-dah-dah-dit. See: **breaker**

L

Lima ("LEE – muh")

Di – dah – di - dit

L – Abbreviation for the electrical term inductance.

ladder line –Balanced, two-conductor transmission line, typically with an impedance between 300 and 600 ohms. The two conductors are separated uniformly by some non-conductive spacing material placed at regular intervals. See: **balanced line, open wire line, window line**

landline (*slang*) (*antiquated term*) – The common telephone. The term is used to make it clear the operator is not talking about the phone mode of transmission. Example: "If we lose each other in the static, give me a call on the landline." See: **twisted pair**

LCD - Liquid crystal display. See: **liquid crystal display**

LDG Electronics – Manufacturer of Amateur Radio equipment. Visit: **www.ldgelectronics.com**

LDE (*slang*) – Long delayed echo. See: **long delayed echo**

lead – (pronounced "leed") 1. A wire or connection point on an electrical component.

2. A probe and the wire attached to it that is used to establish a connection from a test instrument to the point or points in a device where a parameter is going to be measured.

League, The (*slang*) – The American Radio Relay League. See: **American Radio Relay League**

LED - Light-emitting diode. See: **light emitting diode**

LF - Low frequency. See: **low frequency**

Lid (*slang*) – A poor operator.

light emitting diode - A two-lead semiconductor which emits light when activated. Often used as a light source or visual indicator. Abbreviated as LED.

lightning arrestor - A device used to help protect structures, power lines, and equipment from lightning damage by shunting some of the energy to an earth ground system.

linear (*slang*) – An external power amplifier. See: **amp, amplifier, barefoot, linear amplifier**

linear amplifier - An external power amplifier used after the transceiver output. Linear means that the signal emitted by the amplifier is directly proportional to the signal that goes in. See: **amp, amplifier, barefoot, linear**

linear power supply – A device that converts 110-volt alternating current (AC) into nominal 12-volt (typically 13.8 volts) direct current (DC). A steel or iron laminated transformer reduces the input current that is then rectified by diodes and smoothed into low voltage DC by electrolytic capacitors. See: **power supply, switching power supply**

line-of-sight - A form of radio propagation in which the emitted signal travels a straight-line path directly from one station's antenna to the other. This means for good communication, each antenna should be within sight of the other. See: **radio horizon**

liquid crystal display – A type of visual display using two sheets of material with a liquid crystal solution between them. When an electric current is passed through the liquid, the crystals act like a shutter, allowing light to pass through or blocking it. Abbreviated as LCD. Such technology is now often used in Amateur Radio transceiver displays.

little pistol (*slang*) – A modest Amateur Radio station setup with inexpensive, lower-power equipment and limited antennas. See: **big gun, peanut whistle**

LMR-200, LMR-400 - Common types of coaxial cable used by Hams as feedlines for antenna systems. See: **RG-6, RG-8, RG-8X, RG-58, RG-59, 9913**

load - An electrical component or portion of a circuit that consumes power.

lobe – The area in the radiation pattern of an antenna in which the signal strength is at its maximum. See: **null**

log – 1. (*noun*) A document maintained by Hams that lists the details of their stations' contacts as well as other details of operation of the station. Though no longer required by the Federal Communications Commission, such records can come in handy in the case of interference claims and to track propagation conditions. A log is also necessary for claiming a score in most radiosport events.
2. (*verb*) The process of maintaining a log. Many Hams now use computer logging. See: **logging software**

logger (*slang*) – An individual who helps a contest operator by logging contacts and their details.

log periodic antenna – A multi-element beam antenna that is characteristically very broad-banded due to the specifically calculated RF energy interaction among its elements. See: **beam, element**

Logbook of the World – An Internet service on which stations may log contacts for the purpose of applying for the operating awards sponsored by the American Radio Relay League as well as some by *CQ Magazine*. Abbreviated as LOTW. Visit: **http://www.arrl.org/logbook-of-the-world**

logging software – Computer programs that allow Hams to log details of contacts and maintain a database of on-air activity electronically. May be used regularly and/or for radiosport events.

lollipop (*slang*) (*antiquated term*) - Nickname for the distinctively-shaped Astatic D-104 microphone.

long delayed echo (*slang*) – A signal that is heard seconds or even minutes after it was actually transmitted, often by the station that sent it out in the first place, due to the long route it may have taken. The route may have been all the way around the planet. Abbreviated as LDE.

long path – A signal path that is a reciprocal of the shortest route from transmitting station to receiving station. See: **great circle bearing, short path**

loop – An antenna whose radiating element is one continuous conductor, with the feedline connected to each end. The loop is typically hung in a circle, in a square or rectangle, or as a triangle, as with a delta loop. See: **delta loop, horizontal loop, magnetic loop**

LOTW – Logbook of the World. See: **Logbook of the World**

lower side-band, lower sideband - The frequencies on a carrier that are lower than the carrier frequency, but that contain power as a result of the modulation process. Operators using single-sideband can choose to operate either lower or upper sideband. Typically and by convention, LSB is used on 160, 80, 60 and 40 meters. USB (upper sideband) is used on all other bands. Abbreviated as LSB. See: **LSB, upper side-band, USB**

lowest usable frequency - The lowest frequency that can support reliable propagation between stations. Abbreviated as LUF.

low frequency - The radio spectrum between 30 and 300 kilohertz. Abbreviated as LF.

low-pass filter - A filter that allows signals below the cutoff frequency to pass through and reduces signals above the cutoff frequency. See: **high-pass filter**

LSB - Lower side-band. See: **lower sideband**

LTR (*CW abbrev.*) – "Later," "Letter."

LUF – Lowest usable frequency. See: **lowest usable frequency**

LV (*CW abbrev.*) – "Leave," "Love."

LVG (*CW abbrev.*) – "Leaving," "Loving."

M

Mike

Dah – dah

M – 1. Abbreviation for mega or one million. Example: "2 mHz" is two megahertz, or two million cycles per second.

2. Abbreviation for meter, a metric unit of measure of distance. Example: A "20M beam" is a 20-meter beam.

machine (*slang*) - A repeater station. Example: "Let's switch over to the Paducah machine."

MacLoggerDX – A computer software program for logging Amateur Radio contacts and for use in radiosport events. This system is specifically designed for use with Apple computers and the OSX operating system. Visit: **http://www.dogparksoftware.com/MacLoggerDX.html**

magic band (*slang*) – A term used for the 6-meter Amateur Radio band because of its unusual and interesting propagation characteristics.

mag-mount (*slang*) - An antenna with a magnetic base for mounting purposes. This allows quick installation and removal from a motor vehicle or other metal surface. Used primarily for VHF and UHF antennas since such a mount would not be reliable for longer ones.

magnetic loop – A very small but very narrow-banded loop antenna.

magnetic mount – See: **mag-mount**

Maidenhead locator system - A geographic coordinate system of grid squares that divide up maps of the Earth. These grid squares are used by Amateur Radio operators to determine a contact's location. As a personal challenge, many operators attempt to contact stations in as many grid squares as possible. The system's name comes from the English town where it was first created by VHF enthusiasts. See: **grid square**

Main Trading Company – A vendor of Amateur Radio equipment and supplies. Abbreviated as MTC. Visit: **http://www.mtcradio.com/**

maker – One who participates in the "maker movement." Many Hams find themselves right at home in the community of "makers" and many people who are already active in the movement are becoming interested in Amateur Radio. Visit: **http://makerfaire.com/** See: **maker faire**, **maker movement**

maker faire – A gathering of tech enthusiasts, crafters, and hobbyists, who are interested in designing and building things of all types. A part of the "maker" movement. Amateur Radio has become a key element in many maker faires. Visit: **http://makerfaire.com/** See: **maker**, **maker movement**

maker movement - A community of creative and curious people, including hobbyists, enthusiasts and students who are interested in building from scratch all types of things while employing emerging technology. Visit: **http://makerfaire.com/maker-movement/** See: **maker, maker faire**

making the trip (*slang*) - Confirmation that a received signal is readable. Example: "You're making the trip to the repeater just fine today."

malicious interference – Deliberately and intentionally causing interference to another radio transmission. This is not only bad operating practice but illegal and can result in a fine or incarceration.

maritime mobile – An Amateur Radio station operating aboard a maritime vessel. Technically such operation requires that the station be outside the territorial waters of any nation and the operator must follow the rules and regulations of the country under which the vessel is flagged. On CW, /MM should be sent after the call sign of the operator.

MARS – Military Affiliate Radio System. See: **Military Affiliate Radio System**

match (*slang*) – 1. (*noun*) A condition in which the output of a transmitter is the same or nearly the same impedance as the antenna system in use, thus offering the most efficient transfer of power from the rig to the load (antenna). A 50-ohm output to a 50-ohm antenna system is a match. See: **mismatch**

2. (*verb*) The process of finding a point at which the impedance of the antenna system is the same or nearly the same as the output of the transmitter, when capacitive and inductive reactance cancel out leaving primarily radiation resistance and maximum radiation of RF energy. This may already be the case but may also involve using a matching device such as an "antenna tuner." See: **antenna tuner, resonance**

matchbox (*slang*) – A term for an antenna-impedance matching device, commonly called an "antenna tuner." See: **antenna tuner, match**

matching stub - A length of transmission line that is used to help bring an antenna system into resonance. By choosing the proper length and characteristic impedance of line, and having one end open or shorted, a stub becomes in effect a capacitor or inductor and can be used to achieve a match when inserted at a selected point in the regular transmission line. See: **feed line, match, resonance**

maximum usable frequency - The highest radio frequency that can be reliably used for transmission between two points by way of reflection from the ionosphere. Abbreviated as MUF.

mayday – When transmitted over the air, indicates that a life-threatening event is being reported or relayed. This distress call is typically used on voice transmissions as opposed to "SOS" on Morse code or digital modes. See: **distress call, SOS**

MC (*antiquated term*) – Megacycle. One million cycles per second. Now usually replaced by the term "megahertz." See: **hertz**

medium frequency - The radio-frequency spectrum from 300 to 3,000 kilohertz. Sometimes called "medium wave." Abbreviated as MF.

medium wave - The portion of the radio-frequency spectrum from 300 to 3000 kilohertz. Abbreviated as MW. The term is also commonly used to refer to the AM commercial broadcast band, 535 to 1705 kilohertz.

mega - The metric prefix meaning multiply the suffix value by 1,000,000.

megacycle (*antiquated term*) - One million cycles per second, abbreviated MC. Now usually replaced by megahertz. See: **hertz**

megahertz – One million hertz or cycles per second in a recurring wave. See: **hertz**

memories – Programmable settings in a receiver, scanner or transceiver that allow the operator to enter frequencies or channels that are often used so they can be more quickly accessed or scanned. See: **scanner**

memory channel – Frequency, mode, and other information stored by a radio so it can be easily chosen by the operator or scanned by the radio. See: **memories**

memory effect – A phenomenon with certain types of rechargeable batteries in which, if they are not completely discharged before being recharged, they will begin to lose capacity and have to be recharged more often.

meteor scatter – The use of the ionized trails of meteors in or near the Earth's atmosphere to reflect radio signals back to other stations.

Mexican Federation of Radioexperimentadores – The national organization for Amateur Radio Operators in Mexico. Visit at **http://www.fmre.org.mx/**   See: **FMRE**

MF - Medium frequency. See: **medium frequency**

MFJ – A major manufacturer of a wide array of Amateur Radio equipment and accessories, headquartered in Starkville, Mississippi. Visit: **http://www.mfjenterprises.com/**

mic (slang) – Microphone. A device that converts audio into electrical energy. Sometimes spelled "mike."

mickey mouse (*slang*) – Spoken by an operator who is operating maritime mobile, taken from the Morse code identifier for such portable operations, "MM."   See: **maritime mobile**

micro - The metric prefix meaning divide the suffix value by 1,000,000.

microphone – A device that converts audio into electrical energy. See: **mic, mike, PTT**

microphone gain – 1. The sensitivity of a microphone amplifier in a transmitter or transceiver.

2. The control that allows adjustment of a microphone amplifier for more or less gain.

microphone to you (*slang*) – "I have completed my transmission and now it is your turn to talk." Example: "That's all I have for now. Microphone to you, Jack."

microwave – Typically defined as the region of the radio-frequency spectrum above one gigahertz. See: **hertz**

mike (*slang*) – Microphone. See: **microphone**

Military Affiliate Radio System - A civilian auxiliary service sponsored by the Department of Defense in the USA consisting primarily of licensed Amateur Radio Operators who are interested in assisting the military with communications on a local, national, and international basis as an adjunct to normal communications. When participating in MARS activities, stations operate on military frequencies outside the usual Amateur bands. MARS stations must also be licensed by the military in addition to their FCC license and are assigned different call signs than their FCC-issued Amateur call letters. Visit: **http://www.netcom.army.mil/mars**

mill (*antiquated term*) – A typewriter designed to be used by telegraph operators to copy Morse code.

milli - The metric prefix meaning divide the suffix value by 1000.

mismatch (*slang*) - A condition in which the output of a transmitter is not nearly the same impedance as the antenna system in use, thus offering an inefficient transfer of power from the rig to the load. A mismatch that is nearer to the needed impedance may be achieved by using a matching device, sometimes called a "tuner" or matchbox. See: **match, matchbox, tuner**

mixer - A receiver circuit that takes two or more input signals then produces an output that includes the sum and difference of those signals' frequencies. This creates a frequency at which the signal can more easily be processed. See: **heterodyne**

MM – When sent in Morse code following an Amateur Radio call sign, the indication that the station is operating maritime mobile. See: **maritime mobile**

MNI (*CW abbrev.*) – "Many."

mobile – 1. An Amateur Radio station installed in a vehicle or a portable transceiver that can be used in a vehicle.
2. The process of operating an Amateur Radio station from a vehicle of any type. See: **base station, fixed station, portable**
3. (*slang*) Indication by an operator that he is operating mobile. Example: "This is K2XYZ mobile in North Carolina."

mode - Specific type of signal being generated by a transmitting device. Examples: AM, FM, or SSB. Emission modes are more specifically defined by regulatory agencies such as the FCC in the USA. See: **emission mode**

modem - Modulator/demodulator. A circuit or device that modulates an audio or radio signal to transmit data and demodulates a received signal to recover that transmitted data.

modulate – To insert a signal containing information such as audio onto a higher-frequency wave such as a radio-frequency carrier. See: **AM, amplitude modulation, FM, frequency modulation**

modulation - Process by which a signal containing information such as audio is used to modify a higher-frequency wave such as a radio-frequency carrier. See: **AM, amplitude modulation, FM, frequency modulation**

Molex connector - A nylon-body plug often used for power connections.

monitoring (*slang*) – 1. Listening to a radio or scanner. Example: "I'll be monitoring 52 simplex for your call."
2. An on-air indication that an operator is available and listening for anyone to call. Example: "This is K1QQQ monitoring. Anybody around?"

mono-band - A rig, device or antenna that can be used on only one band.

mono-pole – A single element vertical antenna.

moonbounce - See: **Earth-moon-Earth**

MORN (CW abbrev.) – "Morning."

Morse code – A method of transmitting text information as a series of dots and dashes sent and received as on-off tones, lights, or clicks that can be directly understood by a skilled listener or observer. Today, and in Amateur Radio, the International Morse code is the standard version in use. Morse code was named for its inventor, Samuel F.B. Morse. In Amateur Radio, the mode is also often referred to as CW. See the definition for International Morse code for an explanation for why I use dits and dahs rather than dots and dashes at the beginning of each section of this dictionary. See: **CW, code, International Morse code**

MOSFET - Metal-oxide semiconductor field-effect transistor.

motorboating (*slang*) – A fluttering, low-frequency noise on the audio of a transmitting station, so named because the sound resembles that of an outboard boat motor.

mount – The method by which an antenna is attached to a vehicle. See: **ball mount, bumper mount, fender mount, mag mount**

MSG (*CW abbrev.*) – "Message."

MUF - Maximum usable frequency. See: **maximum usable frequency**

muffin fan – A small electrically-powered fan typically used to circulate air through or over electronic equipment.

multi-band – A rig, device or antenna that can be used on more than one Amateur Radio band.

multi-hop - A radio signal that has been refracted by the ionosphere and the Earth two or more times between origination and eventual reception. See: **hop**

multimeter, multi-meter -- A test instrument used to measure various values in a circuit including current, voltage and resistance.

multi-mode transceiver – A transceiver that is capable of operating most emission modes available to Amateur Radio licensees, including single-sideband, CW, AM, and FM.

multi-path – Signals reaching a receiver by way of more than one path due to reflection or refraction off objects, hills, or the ionosphere. This occurs most often on VHF and UHF and can cause a signal to interfere with itself and be difficult to copy. See: **direct path**

multiplier – Criterion within radiosport that allows for greater scoring. Example: The number of individual countries worked in a contest may be multiplied by the total number of contacts, thus making a country a multiplier. If a station makes 500 contacts in 100 different countries in a contest, his score would be 50,000 points (500 X 100).

MW - Medium wave. See: **medium wave**

N

November

Dah – dit

N (*CW abbrev.*) – "No."

N1MM Logger – A popular computer software program for logging Amateur Radio contacts. This program is primarily used for radiosport logging and is available for free. Visit: **http://n1mm.hamdocs.com/tiki-index.php**

N3FJP Amateur Contact Log – An inexpensive and popular commercially available software program for logging Amateur Radio contacts. This program is primarily for day-to-logging use although the author also offers versions of the system for specific radiosport events. Visit: **http://www.n3fjp.com/**

NA (*CW abbrev.*) – "North America."

narrowband FM - A variation of the frequency modulation (FM) mode in which the modulation information on the carrier signal only deviates about 2.5 kilohertz above and below the center frequency. Abbreviated as **NBFM**.

National – A former manufacturer of Amateur Radio equipment, especially known for its line of receivers.

National Contest Journal - A magazine published six times per year by the American Radio Relay League. Covers topics of interest to Amateur Radio operators who enjoy radiosport. Abbreviated as *NCJ*. Visit: **http://ncjweb.com/**

National Electrical Code - A nationally-accepted set of electrical safety guidelines. Abbreviated as NEC. Visit: **http://www.necconnect.org/**

National Institute of Standards and Technology - USA government agency that maintains official standards for the measure of such things as time and frequency. Abbreviated as NIST. Formerly known as The National Bureau of Standards. This agency operates stations WWV and WWVH, broadcasting highly accurate time signals on precise shortwave frequencies. See: **WWV, WWVH**

National Radio Quiet Zone - An area in Maryland, Virginia, and West Virginia near the high-sensitivity radio telescopes operated by various branches of the government. There are varying restrictions on Amateur Radio transmissions—as well as those by other radio services—depending on how close the transmitters are to the protected facilities. Visit: **https://science.nrao.edu/facilities/gbt/interference-protection/nrqz/**

National Traffic System - A national system of Amateur Radio on-air networks designed to relay traffic via Ham Radio in the form of formal messages. Abbreviated as NTS. The NTS is administered by the American Radio Relay League. See: **traffic**

NATO phonetic alphabet – The generally accepted list of words that allows an operator on phone modes to spell out words to distinguish letters for clearer understanding. The heading for each letter in this dictionary shows the phonetic word for that character. See: **phonetic alphabet**

NB - Noise blanker. See: **noise blanker**

NBFM - Narrowband FM. See: **narrowband FM**

NCJ – National Contest Journal. See: **National Contest Journal**

N-connector - A threaded, weatherproof, medium-size RF connector used to join coaxial cables. Preferred use is in VHF and UHF frequency ranges.

NCS - Net control station. See: **net control station**

near field - The region of electromagnetic energy immediately surrounding an antenna.

near-vertical incidence skywave – A form of radio signal propagation in which the emitted wave goes up at a very sharp angle and is reflected back to Earth by the ionosphere in an area within a small radius—depending on the frequency being used—around the transmitting station. Such propagation is best employed for local and regional communication. Abbreviated as NVIS. See: **angle of radiation, critical angle, NVIS**

NEC – 1. National Electrical Code. See: **National Electrical Code**

2. Numerical Electromagnetics Code. See: **Numerical Electromagnetics Code**

negative (*slang*) – "No," "Incorrect."

negative copy (*slang*) – "Unable to copy any of your last transmission or message."

negative offset - A repeater's input frequency is lower than its output frequency. Example: A repeater that transmits at 146.88 megahertz but is listening for users at 146.28 megahertz is said to have a negative offset of 600 kilohertz. See: **positive offset**

net (*slang*) – Network. See: **network**

net control station - The station in control of a formal net who keeps the proceedings orderly and moving along. Abbreviated as NCS. See: **net, network**

network - A group of stations that meets on a specified frequency at a regular time. A net may be informal, formal or somewhere in between. A formal net is organized and directed by a net control station (NCS), who calls the net to order, recognizes stations as they enter and leave the net, and authorizes stations when they may transmit or begin to pass traffic. An informal net may allow participants to transmit at will in no particular order or as a "roundtable," in which stations take turns. Usually shortened to "net." See: **net control station, roundtable, traffic**

next over (*slang*) – The next time the station has a chance to transmit. Example: "John, please tell me your location on your next over."

NiCad - Nickel cadmium, pronounced "NYE – cad." The chemical composition in a popular type of rechargeable battery. The term usually refers to the battery itself. Example: "I have a bunch of NiCads charged and ready to go if we have a power failure."

nickels (*slang*) – A term often used on DX or county-hunting nets meaning a signal report of "55" or "Five by five." Origin: Citizens Band radio. See: **double nickels, RST, signal report**

NIL (CW abbrev.) – "None," "Nothing."

NiMH - Nickel metal hydride. The chemical composition in a popular type of rechargeable battery.

NIST – National Institute of Standards and Technology. See: **National Institute of Standards and Technology**

NMO - A type of screw-on mobile antenna coaxial mounting arrangement often used for VHF and UHF antennas.

node - A remotely controlled digipeater that is used as a connect point in packet radio. See: **digipeater, packet radio**

noise - Undesirable electromagnetic energy that causes interference with a signal's reception.

noise blanker - A circuit designed to reduce pulse-type noise in a receiver. Abbreviated as NB.

noise floor - The total noise in a receiver from all sources, internal and external. To be readable, a signal has to be greater than the noise floor. See: **noise**

noise reduction – A feature of digital signal processing used to reduce unwanted noise on a signal. Abbreviated as NR. See: **digital signal processing, noise**

notch filter – A narrow rejection filter for elimination of interfering signals. This type of filter often can have its rejection frequency varied with a control to be able to reduce an interfering signal wherever it may be above or below the frequency of the desired signal. See: **filter**

Novice (*antiquated term*) – An entry-level Amateur Radio license class in the USA that required a 5-word-per-minute Morse code examination and a basic theory exam. On-air privileges were very limited. Though the license class no longer exists, the Federal Communications Commission continues to renew those licenses that remain in effect.

NR – 1. Noise reduction. See: **noise reduction**
2. (*CW abbrev.*) – "Number."

NTS - National Traffic System. See: **National Traffic System**

null – 1. The area in the radiation pattern of an antenna in which the strength of a radiated signal is its lowest. See: **lobe**
2. To tune or adjust for a negative value.

number stations (*slang*) – Mysterious shortwave stations on which people can be heard reading what appear to be coded messages consisting of all numbers or letters.

Numerical Electromagnetics Code - A general-use software program for modeling antenna design and performance. Abbreviated as NEC. See: **antenna modeling, EZNEC**

NVIS - Near-vertical incidence skywave. See: **near-vertical incidence skywave**

NW (*CW abbrev.*) – "Now."

O

Oscar

Dah – dah – dah

OBS – Official Bulletin Station. See: **Official Bulletin Station**

OC (*CW abbrev.*) – Oceania, the part of the world that includes Australia, New Zealand, and the South Pacific islands.

OCF dipole – Off-center-fed dipole. See: **off-center-fed dipole**

odd split (*slang*) – An unconventional frequency separation between input and output frequencies of a repeater station. Most repeater stations in the USA use a 600 kilohertz split between input and output. See: **offset**, **split**

OES – Official Emergency Station. See: **Official Emergency Station**

off-center-fed dipole - A dipole antenna that is fed at a point away from its center, as with the traditional dipole design. The feedpoint is at a position nearer one end of an element where the impedance will be such that the antenna will work on multiple Amateur bands. Abbreviated OCF dipole. See: **dipole**.

Official Bulletin – A news story or informational article related to Amateur Radio or ARRL activities disseminated through regular multi-mode broadcasts from the American Radio Relay League station, W1AW, via email, by League appointees called Official Bulletin Stations, or other means. The stories are originated by the ARRL. See: **ARRL, American Radio Relay League, Official Bulletin Station, W1AW**

Official Bulletin Station – An appointed volunteer who receives and disseminates to other Hams and the general public Official Bulletins from the ARRL. Abbreviated OBS. See: **ARRL, Official Bulletin**

Official Emergency Station - An appointee in the ARRL's Field Organization who is responsible for specific duties during drills or emergency situations. Abbreviated as OES. See: **ARRL, Field Organization**

Official Observer - A volunteer who monitors the Amateur Radio bands for rules violations or technical problems that may be causing interference and informs the stations involved. Abbreviated OO. The OO program is administered by the American Radio Relay League.

offset (*slang*) – The difference between the input and output frequencies for a repeater station. To be able to receive and transmit at the same time requires two frequencies. In the USA, the standard offset is 600 kilohertz. Usually if the output (transmit) frequency of the repeater is below 147 megahertz the input (listening) frequency is 600 kilohertz lower. This is called a negative offset. If the output is above 147 megahertz, the input is 600 kilohertz above the output. This is a positive offset. See: **negative offset, odd split, positive offset, split**

offset frequency – See: **offset**

ohm - The fundamental unit of measurement of resistance as well as impedance.

Ohm's law – One of the most basic laws of electronics, this is a formula that shows the relationship between voltage (E), current (I) and resistance (R) in an electrical circuit. Ohm's law says that the voltage applied to a circuit is equal to the current flowing through the circuit times the resistance of the circuit, or the mathematical expression: $E = IR$, where E is voltage in volts, I is current in amperes and R is resistance in ohms.

old man (*slang*) – Term used to denote friendship between two male amateur radio operators, regardless their chronological ages. Abbreviated "OM" on CW.

Olivia – An Amateur Radio digital mode designed to work in difficult communication conditions. It is not yet as popular as some other modes and is only available in a few of the commercially available software systems for digital modes. See: **digital modes**

OM (*CW abbrev.*) – "Old man." See: **old man**

omni-directional antenna – An antenna that receives or emits a signal near equally in all directions. See: **directional antenna**

one-by-one call sign - A special temporary call sign consisting of a single letter, a number, and another single letter, issued to special event stations or for other distinct purposes. Example: The call sign N9N was issued for the special event station commemorating the fiftieth anniversary of the journey to the North Pole by the submarine USS *Nautilus*. Visit: **http://www.1x1callsigns.org/**

one-way communication – A transmission on Amateur Radio bands that is not intended to be answered, beyond just simple testing. FCC rules place strict limitations on such transmissions and they are not generally authorized.

OO - Official Observer. See: **Official Observer**

OP (*CW abbrev.*) – "Operator," "Amateur Radio operator," "My name is…" See: **handle**

opening (*slang*) – A condition in which better than usual propagation on a particular band is occurring. Example: "There was a great opening on 6 meters this morning. I worked several Caribbean stations."

open repeater (*slang*) - A repeater whose access is available to any operator who holds license privileges for its operating frequency. See: **closed repeater**

open wire line – A type of balanced antenna feedline that has two conductors spaced uniformly from one end to the other, usually by non-conductive spacers. See: **balanced line, ladder line, window line**

operating event – An on-the-air event for Amateur Radio operators similar to a contest (radiosport) but entrants do not compete against each other for scores. Instead they strive for personal achievement and individual goals. Examples: Field Day, Straight Key Night. See: **contest, Field Day, Straight Key Night**

OPR (*CW abbrev.*) – "Operate," "Operator," "Operating."

OSCAR - Orbiting satellite carrying Amateur Radio. A series of space satellites built, launched and controlled by Ham operators. Visit: **http://www.amsat.org/** See: **AMSAT**

oscillate – To vibrate. In electronics, this typically means to generate an alternating current, radio wave, or other periodic signal.

oscillator – A circuit within a transmitter that initially generates a radio frequency signal.

oscilloscope - an electronic test instrument used to observe wave forms on a display screen.

OT (*CW abbrev.*) (*slang*) – "Old timer." Someone who has been in the hobby for a long time.

outgoing QSL bureau – A system run by the American Radio Relay League to send out QSL cards for Hams in the USA to QSL bureaus in other countries. Cards can only go to countries with QSL bureaus that have a working relationship with the ARRL. Visit: **http://www.arrl.org/outgoing-qsl-service**   See: **QSL, QSL bureau, QSL card**

output frequency - The frequency of a repeater station's transmitter, and the frequency to which a user should have his receiver tuned.

over (*slang*) – 1. When used during a two way communication, it means the transmitting station has completed the transmission and is telling the station to which he or she is talking that it is okay to begin to transmit. Use of the term "over" is not recommended on repeater stations or if propagation conditions are good.
2. An opportunity to make a transmission. Example: "Bob, tell me what kind of rig you are running on your next over."

overload – When a received signal is so strong that it overcomes the normal operating parameters of the receiver's circuits creating undesirable images and false signals at various frequencies.

OVR (*CW abbrev.*) – "Over."

P

Papa ("Puh – PAH")

Di – dah – dah – dit

PA – Power amplifier.

packet - A unit of data sent across a network. In Amateur Radio, packets are sent via the airwaves.

packet cluster - A network of automated Amateur Radio packet stations that communicate information about DX stations that have been spotted on the air, contest reports, and other information.

packet radio - A system of digital communication in which information is transmitted over the air in short bursts. The bursts—or "packets"—usually contain the sending station's call sign along with other information. A terminal node controller (TNC) is used to encode and decode the information for broadcast and reception. Visit: **https://www.tapr.org/** See: **digi-peater, node, terminal node controller**

paddle - a key for sending Morse code with an electronic keyer. Dots to form Morse code characters are made by pressing the paddle one way and dashes by pressing the paddle the other. See: **electronic keyer, keyer**

panadapter – A scope or screen display to visually monitor a wide band of frequencies simultaneously. Sometimes called "spectrum scope." See: **spectrum scope**

parallel conductor feed line – Antenna feed line constructed of two conductors held apart at a constant distance. May be encased in plastic to keep that distance consistent or constructed with insulating spacers placed at intervals along the line.

parasitic – 1. Undesired oscillations in a transmitter that create spurious signals on frequencies other than the desired one.
2. Interaction between elements in a beam antenna. See: **beam, parasitic element**

parasitic element - An antenna element that has effect on how the antenna performs by re-radiating or reflecting RF from the driven element. A parasitic element is not electrically connected directly to the feed line or to any other part of the antenna but, in the case of a beam, is connected mechanically to the boom. See: **beam, boom, parasitic**

Part 97 – The section of the rules and regulations of the Federal Communications Commission that deals with the Amateur Radio Service. See: **Amateur Radio, Amateur Radio Service, FCC, Federal Communications Commission**

pass (*slang*) - The time period when the signals from an orbiting satellite can be heard at a particular location on the ground.

passband - The range of frequencies that will be allowed to pass through a filter without being diminished.

patch (*slang*) – 1. Autopatch. A device that interfaces an Amateur Radio repeater station to the telephone system. This allows a Ham using the repeater to make telephone calls over the air to any telephone. These devices have generally been replaced by cellular telephones. See: **autopatch**

2. (*antiquated term*) A phone patch. See: **phone patch**

PCB – Printed circuit board. See: **printed circuit board**

peak envelope power - The average power sent to the transmission line by the transmitter. Can be calculated by multiplying peak envelope voltage by 0.707. Abbreviated PEP. See: **PEP, PEV**

peak envelope voltage - The maximum voltage on a transmission line while a station is transmitting. Abbreviated as PEV.

peak-reading power meter – A measuring instrument that determines the peak power being run by a transmitter. The result is typically in watts. On some intermittent modes, such as single sideband, the usual standard average-power meter is not mechanically capable of measuring the actual output power. This requires a peak-reading power meter. See: **average power, average power meter**

peanut whistle (*slang*) - A modest Amateur Radio station setup with inexpensive equipment and limited antennas. See: **big gun, little pistol**

pecuniary interest – A reasonable likelihood or expectation that a person will receive cash or something else of value in exchange for a service rendered. Amateur Radio operators are forbidden from having any pecuniary interest in anything he or she does in regard to the hobby. In other words, a Ham may not be remunerated for such activities as relaying a message, reporting an accident, or the like.   See: **Amateur**

pedestrian mobile (*slang*) - An Amateur Radio station operating while walking or running.

PEP - Peak envelope power. See: **peak envelope power**

perigee – A point in the orbit of a satellite in which it passes closest to the earth. See: **apogee**

personal (*slang*) – An operator's first name or preferred on-air name. This term originated as CB jargon but is usually frowned upon by Hams. Most prefer simply "name" or "handle."

PEV - Peak envelope voltage. See: **peak envelope voltage**

phone – An emission mode that uses voice modulation such as AM, SSB or FM.

phone patch (*antiquated term*) (*noun*) 1. A circuit or device that interfaces a telephone with an Amateur Radio transmitter and/or receiver. Such a device could allow a distant Ham to speak to a non-ham on a regular telephone through the interface with a Ham station of someone in the same area as the recipient of the call, avoiding long-distance charges. Phone patches were once common from military bases, persons in remote areas, or ships at sea to family members but are rarely needed now. See: **patch**

2. (*verb*) The process of conducting a conversation using Ham Radio interfaced to a regular telephone. Example: "I used to phone patch a missionary in South America through to his family here in town."

phonetic alphabet – A list of words that allows an operator on phone modes to spell out words to distinguish letters for clearer understanding. The list below is the NATO and International Telecommunications Union recommended version because it is more easily understood by native-speakers in many languages around the world. It is also the one recommended by the ARRL in the USA. That phonetic alphabet is:

- A   Alpha
- B   Bravo
- C   Charlie
- D   Delta
- E   Echo
- F   Foxtrot
- G   Golf
- H   Hotel
- I   India
- J   Juliet
- K   Kilo
- L   Lima (LEE – muh)

M Mike
N November
O Oscar
P Papa (Puh – PAH)
Q Quebec (Kee – BECK)
R Romeo
S Sierra
T Tango
U Uniform
V Victor
W Whiskey
X X-ray
Y Yankee
Z Zulu

picket fencing (*slang*) - A rapidly fluctuating choppy or fluttery signal, usually from a mobile station in motion. The name comes from the sound made by running a stick along a picket fence.

PICON – An acronym meaning "public interest, convenience and necessity." By law and as a trade-off for the creation of the Amateur Radio service, Hams should contribute to public service, assisting their fellow citizens. See: **public service**

pileup, pile-up (*slang*) – A large number of stations all calling a single station at the same time. A pileup usually occurs when a rare DX station is taking calls. See: **DX**

pink ticket (*slang*) – A notice of apparent violation, usually from the Federal Communications Commission, but sometimes also refers to a friendly notice from an Official Observer. See: **Official Observer**

pirate (*slang*) – 1. A station operating illegally by using on the air an existing or random call sign not assigned to him or her.

2. Any station transmitting illegally without possessing a properly issued license.

PKG (*CW abbrev.*) – "Package." "Packing." "Picking."

PL, PL tone - Private Line, a term trademarked by Motorola. Low frequency audio tones on a transmitted signal used to alert or control receiving stations. Same as CTCSS. See: **access code, Continuous Tone Coded Squelch System, CTCSS, tone**

PL-259 – A threaded male connector for coax cable. It is the most commonly used connector for Amateur Radio use and mates with a SO-239 connector. Sometimes referred to as a "UHF connector" though it is not recommended for use at UHF frequencies where the N-connector is a superior choice. See: **N-connector, SO-239, UHF connector**

plug - A male electrical connector designed to be inserted into a jack. See: **jack**

PM (*CW abbrev.*) – "Afternoon," "Evening."

portable – 1. An Amateur Radio station that is designed to be easily moved from place to place but is typically operated from a fixed position, not moving, as in a mobile station. See: **base station, fixed station, mobile**

2. (*slang*) Any operation of a station away from its officially licensed location.

portable designator - Additional identifying information that is added to a call sign to give listeners more information about the station's location. Such a designator typically follows the backslash sign on CW. This is often spoken as "slash" or "portable" on phone modes. Example: "This is N4KC slash maritime mobile." See: **maritime mobile, portable, slash**

positive offset - A repeater station's input (receiving) frequency is higher than its output (transmitting) frequency. Example: A repeater that transmits at 147.14 megahertz but is listening for users at 147.74 megahertz is said to have a positive offset of 600 kilohertz. See: **negative offset**

pot (*slang*) – Potentiometer. An electrical component that is a continuously variable resistor, often used for such purposes as a volume control.

power – The rate at which energy is consumed, expressed in watts. Power in a circuit is calculated by multiplying current by voltage.

Power Poles – See: **Anderson Power Poles**

power supply – In Amateur Radio today this generally means a device that converts 110 volts AC as provided by a normal wall socket in the USA into approximately 13.8 volts DC, the power requirement for most modern radio transceivers. The supply should also be able to handle the current requirements of the radio(s) to which it provides power. See: **linear power supply, switching power supply**

power up – 1. (*adjective*) The condition of a circuit when power has been applied so that it can operate.
2. (*verb*) To switch on the power to a circuit or gear.

PRB – Private Radio Bureau. See: **Private Radio Bureau**

preamp, pre-amp – Preampflier. See: **preampflier**

preamplifier - A circuit in a receiver to boost weak signals before they have been otherwise modified by the receiver. The circuit also increases background noise, but can prove useful if that noise is minimal, as on higher HF and VHF/UHF frequencies. Often shortened to "preamp."

precedence – The priority of a message or traffic communicated via Amateur Radio. This might include 'informal," "formal," "health and welfare," "priority," or "emergency."

prefix – When speaking of an Amateur Radio call sign, the first part of the call including the number. Example: In the call sign WA7XYZ, the prefix is "WA7." See: **call district**, **call letters**, **call sign**, **suffix**

primary service – In cases where two or more radio services are required to share spectrum, the primary service is the one that always has precedence over a secondary service.

printed circuit board - A flat, non-conductive board on which electronic components are mounted and then connected by conductive traces etched ("printed") onto the material. Abbreviated as PCB.

priority channel – On a scanner or a radio with scanning capability, this is a channel that will be automatically selected if a signal is detected there regardless of where the function is in the scan order at the time.

priority traffic – A message or communication via Amateur Radio that is of considerable importance but not of an emergency nature. See: **emergency traffic, informal traffic, net, precedence, traffic**

Private Radio Bureau – The division of the Federal Communications Commission that directly administers Amateur Radio. Abbreviated PRB.

privileges - The frequencies, power levels, modes of communication, and other specific operating parameters that are permitted under the communications rules of a country.

PROB (*CW abbrev.*) – "Problem."

propagation - The means or path by which a radio signal travels from a transmitting station to a receiving station.

prosign – When using Morse code, these are one or two letters sent as a single character to abbreviate longer statements. Examples: "K" for, "Go ahead." "KN" for end of message but no breakers, please." "AR" for end of message. Two-letter prosigns should be sent as a single character. In this dictionary, all prosigns are simply designated as CW abbreviations.

PSE (*CW abbrev.*) – "Please."

PSK31 - Phase shift keying, 31 Baud. A popular radioteletype mode using tones generated from a computer soundcard. The mode is used primarily by amateur radio operators to conduct real-time keyboard-to-keyboard conversations. A variation is PSK63. See: **digital modes**

PTT - Push to talk or press to transmit. Activating a microphone and transmitter by pushing a button somewhere on the microphone, its base, or somewhere else nearby. When the transmission is complete, simply release the button to stop transmitting.

public service – Activities performed by Amateur Radio operators to benefit their communities and fellow citizens. According to law, Hams are to "serve the public interest, convenience and necessity," a phrase reduced to the meme PICON. See: **PICON**

pull the plug, pull the big switch (*slang*) - To shut down the station.

PWR (CW abbrev.) – "Power."

Q

Quebec ("Kee – BECK")

Dah – dah – di – dah

Q – 1. Quality factor. The response of a resonant circuit over a specific bandwidth. Generally a circuit with higher Q is more efficient but has a narrower bandwidth.
2. (*slang*) – A contact or QSO. Example: "Thanks for the Q and see you again soon." See: **QSO**

QCWA - Quarter Century Wireless Association. An organization for Amateur Radio operators who have been licensed for 25 years or more. Visit: **http://www.qcwa.org/**

QEX Magazine - A publication of the American Radio Relay League Published six times a year, the magazine is designed for more technically oriented Hams. Visit: **http://www.arrl.org/qex**

QRL – 1. Q-signal meaning, "Are you busy?" or "I am busy."
2. (*slang*) "Is the frequency in use?" Or, "The frequency is in use." Typically only used on CW or digital modes.

QRM – 1. Q-signal meaning, "Is my transmission being interfered with?" Or, "My transmission is being interfered with." Often used on voice, digital modes and CW.
2. (*slang*) (*noun*) Interference. Example: "There is too much QRM on this frequency for us to carry on."

3. (*slang*) (*verb*) To cause interference. Example: "Let's move. We don't want to QRM the net."

QRN – 1. Q-signal meaning, "Are you troubled by static?" Or, "I am troubled by static." Often used on voice and digital modes as well as CW.
2. (*noun*) Static. Example: "The QRN is especially bad today."

QRO – 1. Q-signal for, "Shall I increase transmitter power?" Or, "I will increase power."
2. (*slang*) High power. Example: "You can tell he is running QRO."

QRP – 1. Q-signal meaning, "Shall I decrease transmitter power?" or "Decrease transmitter power." Often used on voice and digital modes as well as CW.
2. Using very low power while operating, typically five watts or less. Example: "I'm running QRP here today, just a watt output."
3. (*adjective*) Purposely running low-power. Example: "I'm going to compete in the QRP contest this weekend." Or, "They just released their new QRP rig."

QRP ARCI – QRP Amateur Radio Club International. An organization of Amateur Radio operators worldwide who are interested in low-power communication. The group produces a magazine, *QRP Quarterly*), organizes the annual "Four Days in May" conference each year in Dayton, Ohio, and sponsors contests and awards. Visit: **http://www.qrparci.org/** See: **Dayton, Four Days in May**

QRPp (*slang*) – Operating a Ham station while transmitting very low power, typically at one watt or less output. See: **QRP**

QRQ – 1. Q-signal for, "Shall I send (CW) faster?" Or, "Send faster."

2. (*slang*) Operating Morse code at very high speeds. Example: "He is a really good QRQ op."

QRS – 1. Q-signal for, "Shall I send (CW) slower?" Or, "Send slower."

2. (*slang*) – To send CW very slowly. Example: "Most Hams will happily QRS for newcomers if asked."

QRSS (slang) – Operating Morse code (CW) at an extremely low rate of speed, typically at less than one 5-letter character per minute. Often used in marginal propagation, especially VLF and MF. See: **VLF, MF**

QRT – 1. Q-signal meaning, "Shall I stop sending?" Or, "I will stop sending." Often used on voice and digital modes as well as CW.

2. (*slang*) An operator is shutting down his station. Example: "I'm going to have to QRT now."

3. (*slang*) A strong suggestion to another operator to stop transmitting. Example: "Please QRT. You are interfering with emergency traffic."

QRU – 1. Q-signal meaning, "Have you anything for me?" Or, "I have nothing for you."

2. (*slang*) An indication an operator has no more to say and desires to end the contact.

QRV – Q-signal meaning, "Are you ready?" Or, "I am ready." Typically means the receiving station is ready to copy a CW message.

QRX – 1. Q-signal meaning, "When will you call again?" Or, "I will call again in ___ minutes." Often used on voice and digital modes as well.

2. (*slang*) The operator will be standing by for a short time but will return to the air. Example: On CW "QRX 5" means the operator will be standing by for five minutes. Or, "I have a phone call. QRX a minute."

QRZ – Q-signal meaning, "Who is calling me?" Used on phone and digital modes as the equivalent to "CQ." Usually spoken as, "Q R Zed." See: **zed**

QRZ.com, **"QRZ dot com**, **"QRZed dot com"** – A popular Ham Radio web site featuring a "flea market," news, and access to the FCC database for lookup of Ham stations by call sign. Users are also encouraged to upload bio info and photos to the page where their FCC database results are displayed. Often shortened to "QRZ" or "QRZed." Example: "I just looked you up on QRZed and really like your shack." Visit: **www.qrz.com**

QSB – 1. Q-signal meaning, "Are my signals fading?" Or, "Your signals are fading." Often used on voice and digital modes as well as CW.
2. (*slang*) (*noun*) A signal that is fading, or conditions in general in which signals are fading in and out. Example: "It is hard to copy you this morning with all the QSB."

Q-signals - Three-letter codes used by Hams as shortcuts. Though originally created for use in Morse code, many are now commonly used in digital and voice modes as jargon, even if plain language might serve just as well. The more commonly used ones are included alphabetically in this section. When a Q-signal is followed by a question mark in CW, the meaning is in the form of a question. For a complete printable list of Q-signals typically used in Amateur Radio, visit: **http://www.arrl.org/get-on-the-air**

QSK – 1. Q-signal meaning, "Can you work break-in?" Or, "I can work break-in."

2. (*slang*) (*CW abbrev.*) (*noun*) – The capability of employing circuitry when using Morse code (CW) to be able to receive signals between dots and dashes while transmitting. This allows the receiving station to interrupt the communication without waiting for the transmitting station to finish, or for the sending operator to monitor what is happening on the frequency even while transmitting. See: **full breakin**

QSL – 1. Q-signal meaning, "Can you acknowledge receipt?" Or, "I can acknowledge receipt." Often used on voice and digital modes as well as CW.

2. (*slang*) "I understand," or "I copy."

3. (*noun*) Confirmation of a contact. This can be accomplished by sending a card or other written confirmation by regular mail or via several web-based eQSL services. See: **eQSL, LOTW, Logbook of the World**

4. (*noun*) A postcard-like confirmation of a Ham Radio contact, containing details of the contact. Can also be any other form of written confirmation such as a letter.

5. (*verb*) The act of confirming a Ham Radio contact, either by mail or electronically. Example: "Please QSL!" "Yes, I will be happy to QSL."

QSL bureau - Volunteer groups who help stations internationally to exchange QSL cards. They are typically maintained by a country's primary Amateur Radio organization, such as the American Radio Relay League in the USA. Operators are urged to keep the bureau stocked with self-addressed, stamped envelopes.

When a DX operator works a number of stations, he or she will send a batch of cards to the bureau where they are sorted and, when a number of cards have been gathered, sent on to the stations that have envelopes on file. For more on the QSL bureau service for incoming cards, visit: http://www.arrl.org/incoming-qsl-service. See: **bureau, BURO, QSL, QSL card, outgoing QSL bureau**

QSL card - A postcard-like confirmation of a Ham Radio contact. Also a confirmation by shortwave broadcast or commercial broadcast stations for reports by a listener of their signals having been received.

QSL manager - A volunteer who manages the receiving and sending of QSL cards for another Amateur Radio station. Typically the managed station, such as an operator in a rare country, makes so many contacts that he or she would have trouble managing or paying the postage for the volume of incoming QSL card requests.

QSO – 1. Q-signal meaning, "Can you communicate with ____ direct?" or "I can communicate with ____ direct." Often used on voice and digital modes as well.
2. (*slang*) On-air conversation. Example: "Thank you for the nice QSO." Sometimes pronounced "KEW –so."

QSO party (*slang*) – A contest in which stations from a particular state or geographic region attempt to contact as many other operators within or outside their state or region as they can during a set period of time. QSO parties traditionally are not nearly as intense as some other radiosport events. Visit: http://www.hornucopia.com/contestcal/ See: **contest, radiosport**

QST – 1. Q-signal meaning, "Calling all radio amateurs." Typically used at the beginning of a broadcast from W1AW or when an Official Bulletin is about to be relayed by an OBS. See: **Official Bulletin, W1AW**

2. *QST*. The official magazine published by the American Radio Relay League. Available to members in both print and digital versions. Visit: **http://www.arrl.org/qst**

QSY – 1. Q-signal meaning, "Shall I change frequency?" or "I am changing frequency."

2. (*slang*) To change frequency. Often used on voice and digital modes. Example: "Let's QSY to 40 meters and see if the signals are better there."

QTH – 1. Q-signal meaning, "What is your location?" Or, "My location is ____." Often used on voice and digital modes as well as CW.

2. (*slang*) (*noun*) Current location. Example: "My QTH right now is Broad Street in front of the grocery store."

3. (*slang*) (*noun*) City in which the operator resides or the location of the operator's station. Example: "The QTH here is Springfield, Illinois." Though it is unnecessarily wordy, some operators say, "The home QTH is Springfield, Illinois." Or, "I'll sign now. I just got to my home QTH."

quad - A directional wire antenna consisting of two or more one-wavelength loops placed a quarter-wavelength apart. Quad loops can be constructed in a variety of shapes but are usually square.

quagi - An antenna, primarily for VHF and UHF, that employs some of the characteristics and construction of a quad and a Yagi. See: **quad, Yagi**

quarter-wave antenna – An antenna that has a length that is one-quarter-wavelength on the frequency for which it is intended to be used. To work properly, such an antenna requires a counterpoise such as radials, a quarter-wavelength piece of wire, or an automobile body. See: **counterpoise, radials**

quartz crystal - A piezo-electric mineral that can be inserted in an oscillator circuit and cut so it will vibrate at a particular frequency when an electric current passes through it. See: **crystal, crystal oscillator**

question pool - The set of approved questions that are used to put together Amateur Radio license examinations. Each class of license has a separate set of potential questions, and the pool is updated periodically.

Quicksilver Radio – A vendor that sells Amateur Radio equipment and supplies. Visit: **http://www.qsradio.com/**

R

Romeo ("ROH – me – oh")

Di – dah – dit

R – 1) Symbol for the resistance to the flow of electrical current in a circuit, measured in ohms.

2. (*CW abrev.*) "Received as transmitted," "Are."

RAC - Radio Amateurs of Canada. The national Amateur Radio organization for Canadian Hams. Visit: **http://wp.rac.ca/**

RACES - Radio Amateur Civil Emergency Service. A service created and administered by the Federal Emergency Management Agency (FEMA) and the FCC, made up of licensed Hams that are certified by a civil defense agency and are able to communicate on Amateur Radio frequencies during drills, exercises and emergencies. Visit: **http://www.usraces.org/**

RadCom – Official magazine of the Radio Society of Great Britain. Visit: **http://rsgb.org/main/publications-archives/radcom/** See: **RSGB**

radials - Horizontal antenna elements designed to provide an electrical counterpoise to a vertically polarized antenna. Radials may be above ground, on the ground, or buried beneath the ground.

radiation resistance – The quantity of total resistance in an antenna system that causes RF energy to be radiated. This is typically the desired kind of radiation we want. The other kind of resistance in an antenna system is energy that is lost, usually in the form of heat. See: **resonance**

Radio Amateur Satellite Corporation - An educational organization. Its goal is to foster Amateur Radio's participation in space research and communication. The group has been responsible for designing, building and placing into orbit many Amateur Radio satellites. Abbreviated as AMSAT. Visit: **www.amsat.org**

radio check (*slang*) – An on-air query from an operator looking for an evaluation of the station's signal strength and audio quality.

Radio City – A vendor of Amateur Radio equipment and supplies. Visit: **http://www.radioinc.com/**

radio direction finding - Using directional antennas and receivers with calibrated reception metering to locate transmitting stations. Abbreviated as RDF. See: **bunny hunt, fox hunt, radio orienteering**

radio frequency - An alternating current which, if it is input to an antenna, generates an electromagnetic field suitable for such uses as wireless broadcasting. The frequencies of this type of alternating current are generally said to extend from 9 kilohertz to several thousand gigahertz.

radio frequency exposure – The amount, magnitude, and length of time of radio-frequency energy to which any individual is exposed. The Federal Communications Commission has established maximum values that are permitted and Amateur Radio operators must insure their stations meet or exceed those limits, not only for themselves but other persons who are near transmitting equipment and antennas.

radio frequency feedback - Distortion on a voice-modulated signal caused by radio frequency energy finding its way back into the microphone, microphone cable, connector or audio circuit of the transmitter.

radio frequency interference - Interference or spurious noise generated by a source emitting radio-frequency waves. Abbreviated as RFI.

radiogram – A formal piece of message traffic, usually one that is forwarded through traffic nets in the National Traffic System. See: **National Traffic System**

radio horizon - The most distance that radio signals can reliably travel using line-of-sight propagation. See: **line of sight**

Radio Mart – A vendor of Amateur Radio equipment and supplies. Visit: **http://www.radio-mart.net/**

radio orienteering – Using radio direction-finding techniques to determine the location of the listener or transmitting stations. See: **radio direction-finding**

radio shack (*slang*) – The area in which a Ham has set up his radio station. It can be a separate building but is usually just a corner in a room, office, garage, or basement area.

Radio Shack – A chain of stores that sell electronic parts and accessories.

Radio Society of Great Britain - The national organization for Amateur Radio operators in Great Britain. Visit: **http://rsgb.org/**

radiosport (*slang*) – On-air contesting. Though there is a wide variety of contests and competiveness, this typically refers to events in which Amateur Radio operators attempt to contact as many stations as they can within a specified time period in competition with other Hams. The more popular events include the CQ and ARRL DX contests, CQWPX, and Sweepstakes. Visit: **http://www.hornucopia.com/contestcal/** See: **contest, Sweepstakes, WPX**

radiotelephone – Another word for voice-modulated modes of radio communication.

radioteletype – A keyboard/printer mode of communications by radio. At one time, operators used surplus mechanical teletype terminals to enjoy this mode. Today most activity uses computer-generated audio tones or frequency-shift keying and display received text on computer monitors. See: **FSK, frequency shift keying, RTTY**

ragchewing (*slang*) - Chatting informally by way of radio. A "ragchew" is an informal conversation over the air.

random-wire antenna – An antenna consisting of any convenient length of wire, typically connected directly to a transmitter or antenna matching device. Performance of such an antenna can vary greatly.

Raspberry Pi – A very small basic computer that plugs into a TV and a keyboard, yet is capable of many of the things that a desktop PC can do. Typically, it uses the Unix operating system. It can be programmed for use in electronics projects as well as other applications and has found many uses in Amateur Radio. Visit: **https://www.raspberrypi.org/**

RBN – Reverse beacon network. See: **reverse beacon network**

R/C - Radio-controlled. Usually refers to model cars, boats or airplanes that are controlled using radio frequencies.

RCA plug – A type of connector commonly used to carry audio or video signals but often appear as key jacks, auxiliary connectors, or for other uses on Amateur Radio transceivers and equipment. Sometimes called phono or cinch plugs, named for Radio Corporation of America, an early developer. Visit: **https://en.wikipedia.org/wiki/RCA_connector**

RCV (*CW abbrev.*) – "Receive."

RCVR (*CW abbrev.*) – "Receiver."

RDF - Radio direction finding. See: **radio direction finding**

reactance - the opposition to current flow without dissipation of energy such as that of a capacitor or inductor in an AC circuit. Represented by the symbol "X."

reading the mail (*slang*) - Listening to an on-going on-the-air conversation without participating

receiver - A device or circuit that intercepts radio waves and converts them into signals that allow any intelligence carried on those waves to be heard and understood.

receiver incremental tuning - A control on a transceiver that allows the operator to vary the receive frequency a few kilohertz either side of the VFO frequency without affecting the transmitter frequency. Sometimes known as a clarifier. Abbreviated as RIT. See: **clarifier, transmitter incremental tuning**

reciprocal licensing, reciprocal operating authority – Allowing Amateur Radio licensees from foreign countries to operate in the United States without having to acquire a license in the USA. Such permission is based on agreements between the USA and governments of other nations and does not include all countries. Visit: **http://www.arrl.org/cept**) See: **CEPT agreement**

reflected power – Radio frequency power that is reflected back from the feed point of an antenna when there is an impedance mismatch. Non-radiated power may be dissipated as heat when the transmitter is mismatched to the antenna or load, or it may all be eventually radiated, depending on the type of feedline in use. See: **antenna analyzer, forward power, standing wave ratio, SWR**

reflector – The antenna element behind the driven element in a Yagi directional beam. See: **beam, Yagi**

refract – To bend. In relation to radio propagation, this is what the different layers of the ionosphere sometimes do with signals, bending them so they return to Earth at some distant point from their origination.

regulation – How well a power supply controls its voltage output.

relay – 1. A switch that is opened and closed by applying current to an electromagnet.
2. To pass along information from one station to another that is not hearing the first one well enough to copy his call sign and/or message.
3. To assist a net control station by conveying that a station he or she cannot hear is attempting to check into a net. See: **net, net control station**
4. To pass along a piece of formal radio traffic from one station to another. This is how the American Radio Relay League got its name.

remote control – QA situation in which the operator of a station is controlling it from some other location than where the transmitting equipment and antenna are situated. This can be from another room in the home or from the other side of the planet. Such activity has become more popular with the Internet, Skype and other aids. Many Hams now use remote control because of restrictive zoning or covenant limitations on antenna erection.

repeater, repeater station – Radio system, primarily for VHF and UHF frequencies, that receives incoming signals on one frequency and re-transmits them on another. Such stations are typically located at higher elevations and use good receiving antennas since they are intended to extend the range of communications for users' stations. Amateur Radio repeater stations are often sponsored by clubs or organized groups although many individual operators build and maintain them as well.

repeater directory – A listing of repeater stations around the country including their locations, input and output frequencies, open/closed status, and access tones, if any. Such a directory is published by the ARRL. There are also many on-line listings and directory smartphone apps available.

re-set the repeater, re-set the timer - Allowing the time-out circuit for a repeater station to return to zero and once again start timing a new transmission. On most repeaters, this occurs when a station on the input frequency stops transmitting. See: **timer**

resistor – An electrical component designed to offer opposition to the flow of current in a circuit. Its value is measured in ohms.

resonance – The point in a circuit, and especially antenna systems, at which inductive reactance and capacitive reactance are equal and cancel each other out leaving mostly radiation resistance. An antenna does not have to be resonant to perform well but an antenna system—everything from the output of the transmitter to the antenna—should be near resonance in order for the maximum transfer of power to take place. This is an often misunderstood aspect of antenna theory. Visit: **http://www.donkeith.com/n4kc/article.php?p=32**

resonant antenna – An antenna system in which inductive reactance and capacitive reactance are equal and only radiation resistance remains. See: **radiation resistance, resonance**

resonate – (*verb*) To make adjustments in an antenna to achieve a balance of inductive and capacitive reactance so that it is in a state of resonance. What an antenna system does when it is in resonance.

reverse beacon network - A network of mostly automated Amateur Radio stations listening to the bands and reporting stations whose Morse code signals are heard, when and how well. The stations use software called CW Skimmer to detect and report signals heard. Abbreviated as RBN. Visit: **http://www.reversebeacon.net/** See: **CW Skimmer**

RF - Radio frequency. See: **radio frequency**

RF burn - A burn caused by coming into contact with an RF voltage.

RF exposure – Radio frequency energy exposure. See: **radio frequency exposure**

RF feedback – Radio frequency feedback. See: **radio frequency feedback**

RF gain – Usually refers to a control on a receiver that allows the operator to increase or decrease the amount of amplification that is applied to an incoming signal. See: **AF gain**

RF ground – connection of Amateur equipment to Earth ground to eliminate hazards from RF exposure and reduce RFI. See: **RFI**

RFI - Radio frequency interference. See: **radio frequency interference**

RG-6, RG-8, RG-8X, RG-58, RG-59 - Common types of coax cable used as feedlines in antenna systems. See: **LMR-200, LMR-400, 9913**

rice box (*slang*) – Amateur Radio equipment manufactured in China, Japan or elsewhere in the Orient

rig (*slang*) – Equipment used for transmitting and receiving in an Amateur Radio station.

RIG (*CW abbrev.*) (*slang*) Equipment used for transmitting and receiving in an Amateur Radio station.

ripple – A small amount of alternating current remaining after a DC power supply has supposedly rectified and filtered away all AC.

RIT - Receiver incremental tuning. See: **receiver incremental tuning**

R&L Electronics – A vendor of Amateur Radio equipment and supplies. Visit: **http://www.randl.com/shop/catalog/**

rock (*slang*) – Piezo-electric crystal that determines the frequency of a transmitter.

rockbound (*slang*) – Operating with a transmitter that's frequency is controlled by a crystal and thus restricted to one frequency.

roger – "I understand." "I copied your full transmission." Sometimes varied to indicate not full copy, such as, "Roger on most but missed your name."

roger beep (*slang*) - A tone or sound that is heard, usually in Amateur Radio on a repeater station, when a radio operator un-keys his or her microphone. Lets other users know the station is no longer transmitting. Originated in the Citizens Band service where individual stations employed such a tone on their own transmissions. Such practice is not encouraged in Amateur Radio. See: **un-key**

roofing filter - A type of filter found in many modern receivers, typically in the circuit just after the first receiver mixer. Its purpose is to reject stronger signals on adjacent frequencies before more processing of the desired signals take place. See: **filter, mixer**

rotator – An electric motor-driven device attached to an antenna mast to turn so the antenna can be pointed in the desired direction. Many years ago, this device was called a "rotor" and sometimes is today.

rotor (*slang*) (*antiquated term*) – Former word often used to describe a device used to mechanically turn a beam antenna to a selected direction. The proper name is "rotator." See: **rotator**

roundtable (*slang*) – A group of Amateur Radio stations in a conversation, taking turns transmitting. See: **net**

rover (*slang*) – An Amateur Radio mobile station that drives from one spot to another, operating from multiple grid squares, counties, or states over a period of time. Rovers are typically active during county-hunter nets and in contests. See: **county hunter, grid square, radiosport**

RPT (*CW abbrev.*) – "Repeat," or "Report (signal)."

RPRT (*CW abbrev.*) – "Report."

RPTR (*CW abbrev.*) – "Repeater."

RSGB – Radio Society of Great Britain. The national organization for Amateur Radio operators in Great Britain. Visit: **http://rsgb.org/**

RSQ – Readability, signal strength, and quality. A three-digit signal reporting system used to tell an operator how his digital mode signals are being received. Though many digital operators continue to use the well-accepted RST system, RSQ is finding more users because it is more applicable to those modes.

Note that even numbers are not used on the signal-strength and readability scales. The RSQ system is:

READABILITY

1 -- No copy, undecipherable
2 -- 20% copy, occasional words distinguishable
3 -- 40% copy, readable with difficulty, many words missed
4 -- 80% copy, readable with difficulty
5 -- 95% or more copy, perfectly readable

STRENGTH

1 -- Barely perceptible trace
2 -- Not used
3 -- Weak trace
4 -- Not used
5 -- Moderate trace
6 – Not used
7 -- Strong trace
8 -- Not used
9 --Very strong trace

QUALITY

1 -- Splatter over much of the spectrum
2 -- Not used
3 -- Multiple visible pairs
4 -- Not used
5 -- One easily visible pair
6 -- Not used
7 -- One barely visible pair
8 -- Not used
9 -- Clean signal - no visible unwanted sidebars

RST - Readability, signal strength, and tone. The three-digit signal reporting system used to tell an operator how his station is being received. For voice modes, only the first two numbers are used. Tone applies only to CW transmissions.

The RST system is:

READABILITY

1 -- Unreadable
2 -- Barely readable, occasional words distinguishable
3 -- Readable with considerable difficulty
4 -- Readable with practically no difficulty
5 -- Perfectly readable

SIGNAL STRENGTH

1 -- Faint signals, barely perceptible
2 -- Very weak signals
3 -- Weak signals
4 -- Fair signals
5 -- Fairly good signals
6 -- Good signals
7 -- Moderately strong signals
8 -- Strong signals
9 -- Extremely strong signals

TONE

1 -- Sixty cycle AC, or less, very rough and broad
2 -- Very rough AC, very harsh and broad
3 -- Rough AC tone, rectified but not filtered
4 -- Rough note, some trace of filtering
5 -- Filtered rectified AC but strongly ripple-modulated
6 -- Filtered tone, definite trace of ripple modulation
7 -- Near pure tone, trace of ripple modulation
8 -- Near perfect tone, slight trace of modulation
9 -- Perfect tone, no trace of ripple or modulation of any kind

RTTY - Abbreviation for radioteletype. See: **radioteletype**

rubber duck (*slang*) - A shortened flexible antenna usually used with hand-held scanners and transceivers. Not a particularly effective antenna but a trade-off for flexibility and ease of use.

RX (CW abbrev.) – "Receive," "Receiver."

RY – A letter combination sent to test the response of a radioteletype setup. Used because the code that represents these two characters requires the greatest variation in bits from one to the other.

S

Sierra ("See – AIR – uh")

Di – di – dit

SA (*CW abbrev.*) – "South America."

safety interlock -- A switch designed to automatically turn off electricity to a piece of equipment if its cover is removed. This protective device should never be intentionally defeated.

SASE - Self-addressed, stamped envelope. Recommended to be sent with a mailed request for a QSL card. See: **QSL card**

SATERN – Salvation Army Team Emergency Radio Network. A group affiliated with The Salvation Army dedicated to assisting the organization during times of emergency by providing communications when normal means are not available. Visit: **http://www.satern.org/**

scan – Programming a receiver to continually sample designated frequencies or range of frequencies. When a signal is detected, the scan stops temporarily so the operator can listen to the incoming signal.

scanner – A radio receiver with the capability of being programmed to listen to only designated frequencies or a range of frequencies, then to stop temporarily when a station is heard.

schedule (*slang*) - A scheduled on-air conversation at regular time and frequency with another Ham. May be a one-time or a recurring event.

schematic – A drawing or diagram of an electronic circuit.

schematic symbol - A symbol used to represent a component on a schematic diagram. Visit: **https://www.edrawsoft.com/circuit-symbols.php**

School Club Roundup – An operating event sponsored by the American Radio Relay League in which school Amateur Radio clubs attempt to contact each other as well as make contacts with other Hams. Visit: **http://www.arrl.org/school-club-roundup**

screwdriver antenna – A vertical antenna that employs an electric motor to raise and lower the element to attempt to more easily find a match. Some designs actually use motors made for use in electric screwdrivers, thus the name. These antennas are most often used in mobile applications. See: **match, mobile**

SDR – Software defined radio. A radio transmitter and/or receiver in which operations typically done by hardware components (such as amplifiers, detectors, filters, etc.) are accomplished with software, either on an externally attached personal computer or an internal processor dedicated to the radio.

search - A receiver feature that allows the user to set up a frequency range for the receiver to scan. The receiver will then pause on a frequency if a signal is heard.

SEC - Section Emergency Coordinator. See: **Section Emergency Coordinator**

secondary allocation, secondary status – A situation in which a frequency band or portion of a band is shared with one or more other communication services. On a shared band, if the Amateur Radio service is the secondary user, Hams must not cause interference to the primary user.

section – A geographical division of the Field Organization of the American Radio Relay League. Each section has a Section Manager (SM) and other volunteer positions. Sections are often states but in areas with a higher Ham population, the areas are smaller. Example: Georgia and Alaska are sections. Pennsylvania is broken up into Eastern and Western. Visit: **http://www.arrl.org/sections** See: **ARRL**, **Section Manager**, **SM**

Section Emergency Coordinator - A volunteer in the ARRL Field Organization who organizes emergency response by Hams in his or her section. Abbreviated as SEC.

Section Manager - Section Manager. A volunteer in the ARRL Field Organization who organizes Amateur Radio activity in his or her section, appoints Hams to other volunteer positions, and represents the section to the rest of the ARRL's Field Organization.

SED (CW abbrev.) – "Said."

selectivity – The ability of a receiver to reject undesired signals adjacent to the desired signal.

self-spotting (*slang*) – An operator sending his or her own call sign to a DX cluster in an attempt to get other stations to come work the self-spotter. This is generally not encouraged. See: **DX cluster, spotting**

self-supporting – A tower or mast that is designed to be erected and to stay upright without the use of supports or guys. See: **guy**

semi-break-in - Employing circuitry when using Morse code (CW) to be able to receive signals between characters while transmitting. Semi break-in enables an operator to listen to other signals between individual characters and/or words. This allows the receiving station to interrupt the communication without waiting for the transmitting station to finish. This is sometimes referred to as "QSK," from the Q signal for "I can operate break-in." See: **break-in, full break-in, QSK, Q signal**

sensitivity – A measure of how well a receiver can pick up a weak signal.

separation – The frequency difference between a repeater station's input (receive) frequency and output (transmit) frequency. This is sometimes called "split." See: **split**

SET – Simulated Emergency Test. See: **Simulated Emergency Test**

SEZ (CW abbrev.) – "Says."

SFI - Solar flux index. See: **solar flux index**

shack (*slang*) – The area in which a Ham has set up his radio station. It can be a separate building but is usually just a corner in a room, garage, or basement area.

Sherwood rankings – A popular and exhaustive ranking of commercially available Ham Radio receivers and transceivers by Rob Sherwood NCØB. The list is ranked by third order dynamic range, narrow spaced, a criterion that may or may not be of most importance depending on individual user needs. Visit: **http://www.sherweng.com/table.html** See: **dynamic range**

short path - A signal path that is the shortest route from transmitting station to receiving station. The reciprocal path is called long path. See: **great circle route**, **long path**

short skip - Propagation of a radio signal by the ionosphere over a few hundred miles or less.

shortwave – The portion of the radio spectrum usually defined as HF or high frequencies, between 3 and 30 megahertz. See: **high frequency**

shortwave listener – A hobbyist who enjoys listening to the shortwave radio frequencies and international broadcasts. Abbreviated as SWL. See: **shortwave**

sidewinder (*slang*) (*antiquated term*) – An Amateur Radio operator who uses the single-sideband voice mode.

SIG (*CW abbrev.*) – "Signature," "Signal."

sign, sign out, signing off (*slang*) – To end a contact, leave the air and close down the station.

signal – Electrical or electromagnetic impulse that conveys information.

signal generator - A device or circuit that can generate a low-power signal at a designated frequency, typically used for testing purposes.

signal report – An evaluation by a Ham of a another station's readability, signal strength, and technical quality. See: **RSQ, RST, S-meter**

signal strength meter – A meter on a receiver designed to give the relative strength of signals. Usually called an S-meter. See: **S-meter**

SIGS (*CW abbrev.*) – "Signals."

silent key - A deceased Amateur Radio operator. Abbreviated as SK.

simplex - An operating mode in which the transmit and receive frequencies are both the same. See: **duplex**

Simulated Emergency Test - A nationwide emergency communications exercise held annually. Administered by the American Radio Relay League. Abbreviated as SET. See: **Amateur Radio Emergency Service**

single sideband - A voice emission mode in which the carrier is greatly reduced and one sideband of the signal is filtered out. See: **balanced modulator, SSB**

Six Meter International Radio Klub – Organization formed to promote the use of the 6-meter Amateur Radio band. Abbreviated as SMIRK. Visit: **http://www.smirk.org/**

SK (*CW abbrev.*) – 1.) "This is my last transmission." "End of contact." Sent as single character, di-di-di-dah-di-dah.
2. Abbreviation for "Silent key," a deceased Amateur Radio operator.

SKCC – Straight Key Century Club. See: **Straight Key Century Club**

SKED (CW abbrev.) – "Schedule." A scheduled on-air conversation at a regular time and frequency with another Ham. See: **schedule**

skimmer (*slang*) - A multi-channel Morse code (CW) decoder and analyzer program. The system is able to detect and decode most signals heard within a broad range of frequencies and display the call signs of the stations on a web site.

skip (*slang*) – The electromagnetic phenomenon in which signals are reflected or refracted by various layers in the Earth's atmosphere and come back down, sometimes very far away from the original point of origin. The area between those two points, in which the transmission cannot be heard, is called the skip zone. See: **propagation, skip zone**

skip zone (*slang*) – A geographic area in which shortwave signals cannot be heard because the transmitting station is too far away for ground wave propagation and too near for sky wave propagation. See: **ground wave, sky wave**

SKN – Straight Key Night. See: **Straight Key Night**

skyhook (*slang*) – A very large antenna.

SKYWARN – An organization of trained volunteer storm spotters who work with the National Weather Service. Many of its members use Amateur Radio to communicate their reports of dangerous weather. Visit: **http://skywarn.org/**

sky wave – The reflection of radio waves off the ionosphere.

skywire (*slang*) – A large horizontal loop antenna. See: **horizontal loop**

slash (*slang*) – The character: / or "front slash." In Ham Radio, it is typically used to append a portable designator to a call sign. Used on CW as well as phone modes. Example: "This is WA8QQQ slash KH6 in Honolulu."

S-line (*antiquated term*) - A commercially manufactured series of transmitters, receivers, amplifiers, and other Amateur Radio accessories made and sold by the Collins Radio Company from the 1950s to the 1970s.

slim (*slang*) – An unscrupulous person who pretends to be a DX station or other highly desired contact, taking calls and giving signal reports. See: **pirate**

slop bucket (*slang*) (*antiquated term*) – Derogatory term for single sideband, used in the early days of the development of the mode by those who did not like it.

sloper (*slang*) – A dipole or end-fed wire antenna with one end much higher in elevation than the other.

slow-scan television - A picture transmission method to transmit and receive static pictures via radio. This mode is mostly used by Amateur Radio operators on the HF frequency bands. See: **SSTV**

SM - Section Manager. A volunteer in the ARRL Field Organization who organizes Amateur Radio activity in his or her section, appoints Hams to volunteer positions, and represents the section to ARRL headquarters.

SMA – A type of small coaxial cable connector often used to attach an antenna on VHF/UHF portable transceivers.

SMC connector – Sub-miniature C-type. A type of radio-frequency connector typically used for antenna connections to a radio.

S-meter - Signal strength meter. A meter typically part of a receiver that indicates the relative strength of comparable signals. Calibrated in S-units and decibels. See: **decibel, dB, S-unit**

Smith chart - A graphical aid designed to assist in visualizing relationship in and solving problems with transmission lines and matching circuits.

SMIRK – Six Meter International Radio Klub. See: **Six Meter International Radio Klub**

SN (CW abbrev.) – "Soon."

S/N - Signal-to-noise.

S/N ratio – The ratio between a signal and other noise in the receiver. This is one factor to consider in evaluating the performance of a receiver.

SO-239 – A female threaded coax connector, often used on the output of Amateur Radio transceivers. It mates with the PL-259 male connector. See: **PL-259**

soapbox (*slang*) – The section in Ham magazine coverage of radiosport events in which participants talk about their own experiences.

software defined radio - A radio transmitter and/or receiver in which operations typically done by hardware components (such as amplifiers, detectors, filters, etc.) are accomplished with software, either on an externally attached personal computer or an internal processor dedicated to the radio. See: **SDR**

solar flux index - A measurement of solar particles and magnetic fields from our sun that reach our atmosphere, as reported by the Penticton Radio Observatory in British Columbia, Canada. It can vary from values below 50 to values in excess of 300. A higher solar flux index can mean better radio-frequency propagation on the high frequency bands. Abbreviated as **SFI**.

solder – (pronounced "SOD er") 1. (*noun*) A metal alloy that melts at comparatively low heat employed for joining wires or other metals electrically and physically.
2. (*verb*) Join together with solder. See: **soldering gun**

soldering gun, soldering iron – (pronounced "SOD er ing") A device used to melt solder for the purpose of joining two wires or metals together electrically or mechanically. See: **solder**

SOS (*CW abbrev.*) – A distress call that indicates a life-threatening event is being reported or relayed, typically used on Morse code or digital modes as opposed to "Mayday" on voice transmissions. See: **Mayday**

SOTA – Summits on the Air. An award program for Hams and shortwave listeners supporting portable operation in mountainous areas. Visit: **http://www.sota.org.uk/**

SP – 1. (*CW abbrev.*) – "Speaker," "Speed (code)."
2. Abbreviation for "speaker."

spark gap (*antiquated term*) - an early type of transmitter design that employed electrical sparks to generate radio frequency signals.

speaker mic, speaker microphone - An accessory usually used with a hand-held transceiver that contains in one unit both a speaker and a microphone.

special event call sign - A special temporary call sign consisting of a single letter, a number, and another single letter, issued to special event stations or for other distinct purposes. Example: The call sign N9N was issued for the special event station commemorating the fiftieth anniversary of the journey to the North Pole by the submarine USS *Nautilus*. Visit: **http://www.1x1callsigns.org/**

special event station – An operation of an Amateur Radio station designed to acknowledge or commemorate an event, anniversary, festival, etc. Many Ham Radio clubs use such operations to gain public exposure for the hobby. Other Hams enjoy collecting QSL cards or certificates received for contacting such operations.

spectrum – As it pertains to radio, the range of electromagnetic frequencies that are typically used to propagate signals, generally considered to lie between 10 kilohertz and 300,000 megahertz.

Spectrum Monitor – An online e-magazine devoted to hobby radio including Ham Radio. Delivered as a PDF. Visit: **http://www.thespectrummonitor.com/**

spectrum scope – A scope or screen display used to visually monitor a wide band of frequencies at the same time. Sometimes called a "pan adapter." See: **panadapter**

speech processor - A circuit that increases the average level of the audio before it modulates a transmitted signal. Improper adjustment can cause considerable interference to other stations. See: **compression**

splatter - a type of spurious emission that can cause interference to stations on nearby frequencies. Splatter occurs when a transmitter's carrier signal is modulated too heavily. See: **spurious emission**

split – 1. The frequency difference between a repeater station's input (receive) frequency and output (transmit) frequency. This is sometimes called "separation."
2. (*slang*) A DX station operating in such a manner that he is transmitting on one frequency but listening for callers on another nearby frequency. This is to keep a large number of calling stations from covering up the DX station.

sporadic-E – A type of radio signal propagation that occurs when random patches of intense ionization form in the E-layer of the ionosphere and refract higher frequency signals rather than absorb them. This phenomenon typically occurs on 10 meters or above and is especially prevalent on 6 meters and 2 meters in springtime and again around mid-December. See: **E-layer, E-skip**

spots (*slang*) – Listing of DX stations heard or worked on DX spotting web sites. See: **cluster, DX cluster, DX spotting, spotting**

spotting (*slang*) - A process in which stations report hearing ("spotting") or making contact with other stations to a web site where it may be seen by other operators. This allows operators who wish to talk to those stations to go to that frequency and attempt to make the contact. See: **cluster, DX cluster, DX spotting, spots**

sprint (*slang*) – An on-air contest of very limited duration, often as short as an hour or two. See: **contest, radiosport**

spur (*slang*) - Spurious signals. Undesired signals that can come from various sources. They are more serious when present in the output of a transmitter since they can cause interference to other stations. They can also occur in a receiver and make it difficult to hear certain signals.

spurious emission – An undesirable electromagnetic signal that occurs outside the necessary bandwidth of a transmission, such as harmonics, splatter, and the like. See: **harmonic, splatter**

SQL – Abbreviation of "squelch." See: **squelch**

squelch - A circuit that mutes the receiver when no signal or only marginal signals are present, thereby eliminating having to listen to band noise or unreadable signals. Abbreviated as SQL.

squelch tail (*slang*) - A very short bit of noise heard on a repeater station after the end of a radio transmission by a user and before the the receiver's squelch circuit has been reactivated.

SRI (CW abbrev.) – "Sorry."

SSB - Single sideband. A voice emission mode in which the carrier is greatly reduced and one sideband of the signal is filtered out.

SSN - Sunspot number. See: **sunspot number**

SSTV – Slow-scan television. See: **slow-scan television**

standing wave ratio - A ratio between forward and reflected power in an antenna system. It shows how efficiently radio-frequency power is transmitted from a power source (a transmitter or amplifier) through a transmission line (coax, open wire feed line, a single wire), into a load (an antenna). Abbreviated SWR. See: **forward power, reflected power, SWR**

SteppIR – A company that manufactures commercially made antennas for Radio Amateur use. Their primary product is a Yagi-type beam that uses small motors and copper strips to vary the lengths of the beam elements to bring them into resonance. See: **beam, Yagi**

stinger (*slang*) – A small attachment to a whip antenna to make it closer to resonance on a particular band. It typically would contain a coil and a short length of whip.

STN (*CW abbrev.*) – "Station."

straight key (*slang*) - a non-electronic switch mounted on a pedestal used for forming the dots and dashes of Morse code. It typically uses a single spring-loaded lever that is pressed down and released by the operator to close and open a circuit in order to form dots and dashes. See: **key, Morse code**

Straight Key Century Club - An organization of Amateurs who prefer using mechanical keying devices when operating CW. Abbreviated SKCC. Visit: **http://www.skccgroup.com/**

Straight Key Night – Annual operating even sponsored by the American Radio Relay League, held on New Year's Eve/New Year's Day. Operators use CW (Morse code) to hold informal conversations, typically using vintage or otherwise unique equipment and straight keys with which to send the code. Abbreviated as SKN. Visit: **http://www.arrl.org/straight-key-night** See: **ARRL, straight key**

strays – 1. (*slang*) (*antiquated term*) Static.
2. Short informational notices in the pages of *QST Magazine*. See: ***QST Magazine***

strength unit – A unit of signal strength measurement used on S-meters. Abbreviated as S-unit. See: **S-meter**

stub - A length of transmission line that is used to help bring an antenna system into resonance. By choosing the proper length and the characteristic impedance, and having one end open or shorted, a stub becomes in effect a capacitor or inductor and can be used to achieve a match when inserted at a selected point in the regular transmission line. See: **feed line, match, resonance**

subaudible tone, sub-audible tone – A tone that cannot be easily heard on a radio signal because it is below the normal hearing range of humans. Such tones are used for a number of purposes, including controlling access to repeater stations.

suffix - When speaking of an Amateur Radio call sign, the part of the call that follows the number. Example: In the call sign WA7XYZ, the suffix is "XYZ." See: **call district, call letters, call sign, prefix**

S-unit - Markings on the scale of an S-meter, derived from a system of reporting signal strength from S1 to S9 that was developed as part of the RST code. See: **RST, S-meter**

superheterodyne – A type of radio receiver circuit in which an internal signal is generated to mix with a received signal to form a signal at a new frequency. The receiver then processes that new signal, which is now at a frequency that can be more efficiently handled.

SUM (CW abbrev.) – "Some."

Summits on the Air - An award program for Radio Amateurs and shortwave listeners that encourages portable operation in mountainous areas. Visit: **http://www.sota.org.uk/**

sunspot - A storm on the surface of the sun. Such activity can affect the ionization level of the ionosphere. Generally, more spots mean better long distance propagation. See: **sunspot cycle, sunspot number**

sunspot cycle - The regular periods of decline and incline in the numbers of sunspots on the surface of the sun. This is a relatively predictable eleven-year cycle.

sunspot number - An arbitrary numerical value that is used to describe how active the sun's surface is over a period of time. The presence of more sunspots is typically good for radio propagation on the HF portion of the spectrum. Sunspot minima and maxima run in 11-year cycles. Abbreviated as SSN.

superheterodyne - A type of receiver design in which an incoming signal is beat—or heterodyned—with an internally generated signal to create a signal at a frequency that can be more efficiently processed.

surface mount - A method for producing electronic circuits in which the components are mounted or placed directly onto the surface of printed circuit boards. See: **PCB, printed circuit board**

SW 1. (*CW abbrev.*) – "Switch."
2. Abbreviation for "shortwave."

Swan Electronics – A former manufacturer of Amateur Radio equipment.

Sweepstakes – A major radiosport event sponsored each November by the ARRL. Visit: **http://www.arrl.org/sweepstakes** See: **radiosport**

switching power supply - A power supply that uses switching transistors to convert AC to DC rather than a transformer, diodes and electrolytic capacitors like a linear supply. See: **linear power supply, power supply**

switcher (*slang*) – A switching power supply. See: **power supply, switching power supply**

SWL - Shortwave listener. See: **shortwave listener**

SWR - Standing wave ratio. See: **standing wave ratio**

SWR meter – A measuring device used to determine the standing wave ratio of an antenna system.

T

Tango

Dah

T (*CW abbrev.*) – Zero ("dah"). See: **cut numbers**

talk-in (*slang*) – Giving driving directions to incoming visitors to a Ham Radio event or venue, most often on a VHF or UHF repeater.

talk-in frequency (*slang*) – The designated frequencies or channels on which operators offering driving directions to a Ham Radio event or venue can be found.

talk out (*slang*) – To talk longer than the amount of time at which a repeater station's talk out timer is set and have the repeater stop transmitting. See: **courtesy beep, talk out timer**

talk out timer (*slang*) – A circuit in a repeater station that will turn off the repeater's carrier if a user talks longer than the amount of time for which it is set. This is to prevent someone from dominating the repeater and also to allow the transmitter's safe duty cycle to be maintained. See: **courtesy beep, duty cycle, talk out**

TAPR - Tucson Amateur Packet Radio Corp. An organization that supports research and development in Amateur Radio digital communications modes. Visit: **https://www.tapr.org/** See: **packet radio**

Technician, Technician class – The currently available introductory class of Amateur Radio license in the USA. See: **Amateur Extra, General**

telegraphy - The transmission of information in Morse code. Pronounced "tuh LEG ruh fee." See **CW, Morse code**

telemetry - A one-way transmission using radio that carries information that might be used for tracking and to carry measurement data.

telephony - The transmission of information in a voice mode. Pronounced "tuh LEF oh nee." See: **AM, FM, SSB**

telescoping antenna - An antenna that slides into and out of itself to become shorter or longer for storage or matching purposes.

temporary state of emergency – A declaration by the Federal Communications Commission that an emergency has occurred in a designated area. Specific rules then apply to all Amateur Radio operators in that area for the duration of the emergency.

Ten Tec – A major manufacturer of Amateur Radio equipment. Visit: **http://www.rkrdesignsllc.com/**

Ten Ten International - An organization of Amateur Radio operators dedicated to maintaining interest in operating on the 10-meter Ham band. Abbreviated as 10-10. Visit: **http://www.ten-ten.org/**

terminal node controller – A device used as a modem to encode and decode packets of digital data onto a signal for use in packet radio. Abbreviated as TNC. Visit: **https://www.tapr.org/** See: **modem, packet radio**

TEST (*slang*) (*CW abbrev.*) – On-air contest, radiosport event. The term may also be used on voice modes, too. Example: "CQ TEST CQ TEST DE PJ1XX."

test session – An event in which the Amateur Radio license examination is administered by Volunteer Examiners. See: **Volunteer Examiner**

Texas Towers – A vendor of Amateur Radio equipment and supplies. Visit: **http://www.texastowers.com/**

The League (*slang*) - The American Radio Relay League. See: **American Radio Relay League, ARRL**

third-party – An unlicensed person for whom traffic might be passed via Amateur Radio. See: **third party communications, third party traffic**

third-party communications, third-party traffic - Messages passed from one Amateur Radio operator to another on behalf of a third person. Amateurs may not receive any payment for handling such messages.

third-party communications agreement - An official understanding, confirmed by treaty, between one country and another allowing licensed Amateur Radio operators in both countries to pass along third-party communications from one to the other. See: **third-party communications**

third-party operation – Operation of an Amateur Radio station by an unlicensed person. Such operation must be under the direction of a control operator.

THRU (*CW abbrev.*) – "Through." "Threw."

ticket (*slang*) - An Amateur Radio license.

TIL (*CW abbrev.*) – "Until."

time out (*slang*) - To talk longer than the amount of time at which a repeater station's talk out timer is set and have the repeater stop transmitting. See: **courtesy beep, talk out timer**

timer (*slang*) - A circuit in a repeater station that is set to shut off the repeater's carrier if a user talks longer than the amount of time for which it is set. This prevents someone from dominating the repeater and allows the transmitter's duty cycle to be observed. See: **courtesy beep, talk out**

TKS (*CW abbrev.*) – "Thanks."

TMW (*CW abbrev.*) – "Tomorrow."

TMRW (*CW abbrev.*) – "Tomorrow."

TNC – Terminal node controller. See: **terminal node controller**

TNX (*CW abbrev.*) – "Thanks."

tone - 1. (*slang*) A sub-audible tone needed for computer station access See: **access code**

2. An audio-frequency signal generated at a pre-determined constant or varying frequency used to test modulation, signal waveform, power output, and other parameters.

tone access – The necessity for a sub-audible tone of a specific frequency that must be inserted into an operator's audio in order for a repeater station to re-broadcast the incoming signal. See: **access code, tone**

tone pad – A device with twelve or sixteen numbered touch keys each of which can generate a standard telephone multi-frequency two-tone dialing signal. Tone pads are typically built into hand microphones that are supplied with VHF/UHF transceivers since they are often used to perform various functions on a repeater station. See: **autopatch**

top band (*slang*) – the 160-meter Ham band, so called because it has always been (and is as of this writing) the highest assigned frequency band when considering wavelength.

toroid - A donut-shaped device usually made of ferrous materials that can be used as an inductor, such as on coax feedlines, to choke off unwanted RF energy that might travel the outside of the cable's shield.

T/R - Transmit/receive.

TR (*CW abbrev.*) – "Transmit," "Transmitter."

traffic (*slang*) - A message or messages sent by radio. Such communication can be formal or informal. See: **emergency traffic, informal traffic, formal traffic, National Traffic System, net, precedence, NTS**

transceiver – A communication device that is capable of both transmitting and receiving radio-frequency signals, usually employing much of the same circuitry to perform both processes.

transducer (*antiquated term*) – An early word for antenna.

transient - A very short burst of energy on a power line, usually lasting for fractions of a second. Transient spikes can still cause damage to equipment connected to the power line.

transmatch – Another word for "antenna tuner." See: **antenna tuner, matchbox**

transmission line – 1. A term for feedline. See: **feedline**
2. A set of wires for carrying commercial power cross country.

transmitter - A circuit that generates radio-frequency signals of enough strength to allow for communication. Abbreviated as XMTR.

transverter - A circuit in or attached to a transceiver or transmitter that allows the radio to operate on other bands. This capability is typically employed so a transceiver designed only for HF can be used on VHF or UHF bands. Abbreviated as XVTR.

TRBL (*CW abbrev.*) – " Trouble."

tropo (*slang*) – Term for tropospheric ducting propagation. Pronounced "TROP oh." See: **tropospheric ducting**

tropospheric ducting – The propagation of radio signals via bending and ducting along weather fronts. Such phenomena occur in the lowest layer of the Earth's atmosphere, the troposphere, and is most present on frequencies above 30 megahertz. See: **propagation**

T/R switch – A switch or relay used to change an antenna feedline from the transmitter output to the receiver input.

trunked radio system - A computer-controlled scheme used in two-way radio communications that allows more efficient use of relatively few radio frequency channels while helping prevent eavesdropping. Abbreviated as TRS.

TRX (*CW abbrev.*) – "Transceiver," "Transmitting."

TT (*CW abbrev.*) – "That."

TU (*CW abbrev.*) – "Thank you."

tube – Vacuum tube. See: **vacuum tube**

tuner (*slang*) – Matchbox, antenna matching device, "antenna tuner."

TVI (*slang*) – Interference to a television set by an Amateur Radio station's transmissions. Also sometimes used in regard to interference to Amateur Radio receivers by a television set.

twin-lead – A type of balanced twin-conductor feedline encased in plastic that keeps the two wires an equal distance apart. Twin-lead was once used extensively for TV antenna feedline.

twisted pair (*slang*) (*antiquated term*) – The common telephone. The term is used to make it clear the operator is not talking about the phone mode of radio transmission. Example: "If we lose each other in the static, give me a call on the twisted pair." See: **landline**

two-tone test - A method of testing the audio or power output of a single sideband transmitter by feeding two audio tones of different frequencies into the microphone input of the transmitter and observing the output on an oscilloscope. The two simultaneous tones more closely represent the tonal characteristics of the human voice.

TX (*CW abbrev.*) – "Transmitter," "Transmit." See: **transmitter**

TXT (*CW abbrev.*) – "Text," typically referring to the part of a formal piece of radio traffic that contains the content of the message.

U

Uniform

Di – di – dah

U (*CW abbrev.*) – "You."

UHF – Ultra high frequency. See: **ultra high frequency**

UHF connector – See: **PL-259**

ULS – Universal Licensing System. See: **Universal Licensing System**

ultra high frequency - The portion of the radio-frequency spectrum between 300 and 3000 megahertz. Abbreviated as UHF. See: **EHF, HF, VHF, VLF**

unbalanced line - A feedline in which one conductor, the shield, is designed to be at ground potential. The most common example in Amateur Radio is coax cable.

Uncle Charlie (*slang*) - The Federal Communications Commission.

uncoordinated repeater – An Amateur Radio repeater station operating without the approval of any recognized frequency coordinating group.

uninterruptable power supply - A battery back-up power system that can provide alternating current in the event of loss of commercial power. Most modern versions also include protection for equipment from power line transient spikes as well. Abbreviated as UPS.

Universal Licensing System - The Federal Communications Commission's database and systems for processing application filings and more for all wireless services in the USA, including Amateur Radio. Abbreviated as ULS.

Universal Radio – A vendor of Ham Radio equipment and supplies. Visit: **http://www.universal-radio.com/**

unkey, un-key (*slang*) – To release the push-to-talk switch on a microphone to cease transmitting. See: **key, key-up**

unun – A component designed to couple an unbalanced antenna of one impedance to an unbalanced feed line of a different impedance with as small a mismatch as possible. The typical ratios include 1-to-1, 4-to-1, and 9-to-1. Example: A feedline has an impedance of 200 ohms and we want to feed a 50-ohm dipole with it. A 4-to-1 unun would be chosen for the job. See: **balun, match**

UP (*CW abbrev.*) (*slang*) – "Listening up in frequency." An indication by a highly sought DX station that the operator will be listening up in frequency for calls. This keeps the large number of callers from interfering with the DX station on his transmitting frequency. Example: "I'm listening up five." See: **DN, down, DWN**

uplink - The frequency on which a user transmits to a repeater station or satellite. See: **downlink**

upper side-band, upper sideband - The frequencies on a carrier that are higher than the carrier frequency, but that contain power as a result of the modulation process. Operators using single-sideband can choose to operate either lower or upper sideband. Typically and by convention, LSB is used on 160, 80, 60 and 40 meters. USB (upper sideband) is used on all other bands. Abbreviated as USB. See: **lower side-band**, **LSB**, **USB**

UPS – Uninterruptible power supply. See: **Uninterruptible power supply**

UR (*CW abbrev.*) – "Your," "You're."

URS (*CW abbrev.*) – "Yours."

USB – 1. Upper side-band. See: **upper side-band**
2. Universal Serial Bus. Cables, connectors and communications protocols for connection, communication, and power supply between computers and other electronic devices, such as Amateur Radio transceivers.

UTC - Coordinated Universal Time. See: **Coordinated Universal Time**, **Greenwich Mean Time**, **Zulu time**

V

Victor

Di – di – di - dah

V – A common symbol for volt, a unit of electromotive force {EMF}.

vacuum tube – An electronic device consisting of a system of electrodes arranged in glass or metal envelope from which most of the air or other gas has been removed. Such devices were once commonly used in radios, televisions and other components but have been mostly supplanted by the transistor except in high-power amplification uses.

vacuum tube voltmeter - A voltmeter employing vacuum tubes. Abbreviated as VTVM. This type meter is helpful because it has a very high input impedance. This makes a VTVM valuable for measurements in circuits from which only very small currents can be drawn without altering the voltages being measured.

valve (*antiquated term*) – A vacuum tube. The term is still used in Europe and especially Great Britain.

vanity call - An Amateur Radio call sign that has been specifically requested by and issued to the operator who holds it. With some exceptions, a call sign that is currently not issued to anyone may be requested, but certain combinations are not available to all classes of license. The vanity program is administered by the Federal Communications Commission. Visit: **http://www.arrl.org/vanity-call-signs** See: **call letters**

variable frequency oscillator - A circuit within or attached to a transmitter or transceiver that varies the frequency on which the radio transmits. This lets the operator move within the bands covered by the radio and for which he or she is licensed. Abbreviated as VFO.

VE – 1. Volunteer Examiner. See: **Volunteer Examiner**
2. The call sign prefix for many Canadian Amateur Radio stations.
3. An Amateur Radio operator from Canada. Most Canadian call signs begin with the letters "VE." Example: "I had a nice chat on 40 meters with a 'VE' from Toronto."

VEC - Volunteer Examiner Coordinator. See: **Volunteer Examiner Coordinator**

velocity factor - The speed at which radio-frequency energy travels through particular conductors or feedline designs, expressed as a percent of the speed of light through a vacuum.

vertical antenna – An antenna with a radiating element that is vertical to the ground or other surroundings, and usually with radial elements arrayed beneath it in a spoke pattern.

vertical polarization – 1. An electromagnetic wave that has its electrical lines of force perpendicular to the ground.

2. A state in which an antenna has its element or elements perpendicular to the ground beneath it. See: **horizontal polarization**

very high frequency – The portion of the radio frequency spectrum between 30 and 300 megahertz. Abbreviated as VHF. See: **EHF, HF, VLF, UHF**

very low frequency - The range of radio frequencies lower than 30 kilohertz. Characterized by very long wavelengths. Long used for military communications with submerged submarines. Abbreviated as VLF. See: **EHF, HF, UHF, VHF**

VFO - Variable frequency oscillator. See: **variable frequency oscillator**

VHF - Very high frequency. 30-300 MHz-range signals. See: **very high frequency**

Visalia (*slang*) – Commonly used name for the International DX Convention held annually in Visalia, California. Alternately sponsored by Northern California DX Club and Southern California DX Club, the event is primarily for DXers and contesters. Visit: **http://www.dxconvention.com/**

VLF – Very low frequency. See: **very low frequency**

VOA – Voice of America. See: **Voice of America**

voice keyer - A device that can store and transmit pre-recorded voice messages. Especially useful in radiosport or for calling CQ. See: **CQ, radiosport**

Voice of America - A group of high-powered international shortwave stations operated by the State Department of the USA. Abbreviated as VOA.

voice operated relay - A circuit or component that senses the presence of sound through a microphone and turns on the carrier in a transmitter/transceiver to transmit. When there is a pause, the relay opens and the carrier is dropped. Usually abbreviated as VOX and pronounced "vocks." It is most often used in single sideband (SSB) operation. See: **anti-VOX**

voltage - The amount of difference in potential energy between two points in an electrical circuit. Sometimes called electromotive force or EMF. The unit of measurement of voltage is the volt. Usually abbreviated as E. See: **EMF**

voltage standing wave ratio – Abbreviated VSWR. Virtually the same as standing wave ratio (SWR). See: **standing wave ratio, SWR**

voltmeter – A device for measuring the voltage between two points in a circuit.

volt-ohm meter - A test instrument that can be used to measure current, voltage, resistance, electrical continuity, and other parameters in an electrical circuit. Abbreviated as VOM.

Volunteer Examiner - A person authorized to administer Amateur Radio license examinations. Abbreviated VE. See: **exam session**

Volunteer Examiner Coordinator - An Amateur Radio organization empowered by the Federal Communications Commission to recruit, organize, regulate and coordinate Volunteer Examiners and to assure the integrity of Ham Radio license examination sessions. Abbreviated VEC. See: **VE, VEC, Volunteer Examiner**

VOM - Volt-ohm meter. See: **volt-ohm meter**

VOX - Voice operated relay. See: **voice operated relay**

VSWR – Virtually the same as standing wave ratio (SWR), though it is specifically referring to the ratio between the highest voltage along a transmission line to the lowest voltage found at another point on the same line. Most simply refer to this value as SWR. See: **standing wave ratio, SWR**

VTVM - Vacuum tube voltmeter. See: **vacuum tube voltmeter**

VY (*CW abbrev.*) – "Very."

W

Whiskey

Di – dah – dah

W1AW – The Amateur Radio station of the American Radio Relay League in Newington, Connecticut. In addition to allowing for regular operating by staff and visitors, the station also broadcasts Official Bulletins, runs regular code practice sessions, and more. The call sign originally belonged to one of the founders of ARRL, Hiram Percy Maxim. See: **ARRL, Official Bulletins**

W4RT Electronics – A vendor of Amateur Radio equipment and supplies. Visit: **http://www.w4rt.com/**

WAC - Worked All Continents. An operating award for making confirmed contact with stations on each of the world's continents. Administered in the USA by the American Radio Relay League.

walkie-talkie (*slang*) - A small, hand-held, battery-powered transceiver, usually for the VHF and/or UHF frequencies. Sometimes called a handi-talkie, HT, or brick. See: **brick, handie-talkie, HT**

wallpaper (*slang*) - QSL cards, awards and special event certificates, and other items an Amateur Radio operator might proudly hang on his wall for others to see.

wall wart (*slang*) - A small switching power supply unit for low-power equipment that plugs into a standard AC wall outlet. These devices are also notorious for generating electrical noise and interference.

WAN (*slang*) – "Worked all neighbors." See: **worked all neighbors**

WARC - World Administrative Radio Conference. See: **World Administrative Radio Conference, ITU**

WARC bands (*slang*) - An expression to indicate the additional three bands that were allocated to the Amateur Radio service in most of the world at the WARC conference in 1979. Those bands are 12, 17, and 30 meters. Note that by gentlemen's agreement, no contests/radiosport events are held on these three bands. See: **World Administrative Radio Conference**

WAS - Worked All States. See: **Worked All States**

waterfall (*slang*) - A display used with digital modes that is made up of horizontal lines moving down the computer monitor screen, resembling a waterfall.

watt – The unit of measurement of power.

wavelength - The distance between successive crests of a wave, and especially between the same points in sound waves or electromagnetic waves. Typically, in the case of radio (electromagnetic) waves, the unit of measurement is meters.

WAZ – Worked All Zones. See: Worked All Zones

WBØW – A vendor of Amateur Radio equipment and supplies. Visit: **http://www.wb0w.com/**

Weak Signal Propagation Reporter – Abbreviated WSPR. Pronounced "whisper." A computer program used for digital over-the-air communication between Amateur Radio stations. The program is especially helpful with low-power transmissions over difficult propagation paths on the MF and HF bands. See: **digital modes, HF, MF**

WEFAX - Weather facsimile, satellite images and photographs transmitted by government weather satellites in orbit. Many Hams enjoy finding and downloading this data.

wet shoestring (*slang*) – Derogatory term for a very poor antenna.

WFWL (*CW abbrev.*) (*slang*) – "Work first, worry later." An expression used by DX chasers when the validity or the true identity of a station is in doubt. See: **pirate, slim**

whip antenna (*slang*) – A vertical antenna whose single element is a flexible rod or tube. Sometimes called a "stinger," and sometimes has an attachment to adjust its resonant frequency that is called a "stinger." See: **stinger**

WIA – Wireless Institute of Australia. The national Amateur Radio organization of Australia. Visit: **http://www.wia.org.au/**

wide-range antenna tuner - A matching device between transmitter and feedline that can compensate for very large impedance mismatches. See: **matchbox, tuner**

wilco (*slang*) (*antiquated term*) – Literally "I will comply." Once commonly used as, "Roger wilco," meaning "I understand and will comply."

windom antenna – An off-center-fed dipole antenna designed to present a reasonable match on several Amateur Radio bands. See: **OCF, off-center fed antenna**

window (*slang*) - A range of frequencies set aside by gentlemen's agreement to allow for foreign Amateur Radio stations to call CQ and work other stations around the world while those in the United States and Canada refrain from other types of activity there. The DX stations may also invite U.S. and Canadian stations to call them. Sometimes called a "DX window." Example: 3.790 – 3.800 megahertz is the 75 meter DX window. Visit: **http://www.bandplans.com/**

window line - A type of antenna feedline using two parallel conductors encased in plastic insulation. Holes have been punched in the middle area between the two conductors to reduce weight and so the line does not fluctuate so much in the wind. See: **balanced line, ladder line, open wire line**

Winlog32 – A popular software program for logging Amateur Radio on-air contacts and for use in radiosport events. The program is available for free download. Visit: **http://www.winlog32.co.uk/**

wireless (*antiquated term*) – A term once used to identify radio communications as opposed to wired means such as by telegraph.

Wireless Institute of Australia – The National Amateur Radio organization of Australia. Abbreviated as WIA. Visit: **http://www.wia.org.au/**

work (*slang*) - To carry on a valid two-way radio contact with another Amateur Radio station. Example: "I see in the log that we worked back in 2007."

Worked All Continents - An operating award for making confirmed contact with stations on each of the world's continents. Administered in the USA by the ARRL. Visit: **http://www.arrl.org/wac**

worked all neighbors (*slang*) – A fictitious award for a Ham Radio operator with a serious TVI or RFI problem, meaning he has interfered with all the people living nearby. Abbreviated as WAN. See: **RFI, TVI**

Worked All States - An operating award for making and confirming contact with Ham Radio operators in each of the fifty United States. Administered by the American Radio Relay League. Abbreviated as WAS. Visit: **http://www.arrl.org/was**

Worked All Zones - An operating award for making and confirming contacts with other licensed Amateur Radio operators from each of 40 zones, geographical areas designated by *CQ Magazine*, which administers the award. Visit: **http://www.cq-amateur-radio.com/cq_awards/** See: *CQ Magazine*, **CQ zones**

World Administrative Radio Conference - A technical conference of the International Telecommunication Union (ITU) where delegates from member nations of the ITU meet to revise or amend the international radio communications treaties and agreements. A major part of the conferences is determining assignment of parts of the radio spectrum to the various parties who use them. Abbreviated as WARC. See: **ITU**

World Radio Laboratories – A former manufacturer of Amateur Radio equipment, best known for its line of Globe transmitters. Abbreviated as WRL.

Worldwide Radio Operators Foundation - An independent organization devoted to the skill and art of radio operating. The group focuses on operating with emphasis on radiosport and how such activities can improve communications ability and station competency. Abbreviated as WWROF. Visit: **http://wwrof.org/**

worm burner (*slang*) - An antenna system so near to the ground that it tends to have most of its energy absorbed by the earth beneath it. See: **cloud warmer**

wouff hong (*slang*) – A mysterious and highly dreaded weapon that was supposed to be used on any Ham who insisted on exhibiting poor operating procedures on the air. Created by Hiram Percy Maxim, the original holder of the W1AW call sign and a co-founder of the American Radio Relay League. Visit: **http://www.arrl.org/ham-radio-history**

WPM (*CW abbrev.*) - Words per minute. How fast Morse code is being sent or received. A "word" is defined as five characters. See: **Morse code**

WPX – World prefix contest. A radiosport event in which operators attempt to work as many stations with unique call sign prefixes as possible within a set time period, typically a weekend each year for phone modes and a weekend for CW. Sponsored by *CQ Magazine*. Visit: **http://www.cqwpx.com/** See: **prefix, radiosport, suffix**

WRK (*CW abbrev.*) – "Work."

WRKG (*CW abbrev.*) – "Working."

WSJT - A computer program used for weak signal communications between Ham Radio operators. See: **digital, digital modes, JT-65**

WSPR - Weak signal propagation reporter. Pronounced "whisper." A computer program used for digital communication between Amateur Radio stations. The program is especially helpful in sending and receiving low-power transmissions over difficult propagation paths on the MF and HF bands. See: **digital modes, HF, MF**

WWROF – Worldwide Radio Operators Foundation. See: **Worldwide Radio Operators Foundation**

WWV – A radio station operated by the National Institute of Standards and Technology, an agency of the U.S. Department of Commerce. The station offers a frequency reference source by allowing anyone to calibrate to their transmit frequency. They also send highly accurate time signals and radio propagation reports on 2.5, 5, 10, 15 and 20 megahertz, and at times on 25 megahertz, all from the location in Ft. Collins, Colorado. Visit: **http://www.nist.gov/pml/div688/grp40/wwv.cfm** See: **WWVB, WWVH**

WWVB – A radio station operated by the National Institute of Standards and Technology in Ft. Collins, Colorado. The station broadcasts time signals that are accessed by radio clocks all over the world, which synchronize to WWVB's highly accurate telemetry. Visit: **http://www.nist.gov/pml/div688/grp40/wwvb.cfm**

WWVH – A sister station to WWV, WWVH is the U.S. National Institute of Standards and Technology's shortwave radio time signal station. It broadcasts from the island of Kauai in the state of Hawaii. Visit: **http://tf.nist.gov/stations/wwvh.htm** See: **WWV**

WX (*CW abbrev.*) – "Weather."

WX4NHC – The Amateur Radio station at the National Hurricane Center in Miami, Florida. Visit: **http://w4ehw.fiu.edu/**

X-ray

Dah – di – di – dah

X – Symbol for the electrical property known as reactance.

XCVR (*CW abbrev.*) – "Transceiver."

XIT - Transmit incremental tuning. A control that allows the operator of a transceiver to change the transmit frequency while leaving the receiver on its original frequency. See: **RIT, receiver incremental tuning**

XLR connector – A type of audio connector typically found on professional audio equipment.

XMIT (*CW abbrev.*) – "Transmit."

XMTR (*CW abbrev.*) – "Transmitter."

XTAL (*CW abbrev.*) – "Crystal." See: **crystal**

XVTR – Transverter. A circuit in or attached to a transceiver or transmitter that allows the radio to operate on other bands. This capability is typically employed so a transceiver designed only for HF can be used on VHF or UHF bands.

XYL (*CW abbrev.*) (*slang*) – "Wife." "Ex-young lady."

Y

Yankee

Dah – di – dah – dah

Yaesu – A major manufacturer of Amateur Radio equipment, headquartered in Japan. Visit: **https://www.yaesu.com/**

Yagi - A beam-type directional antenna consisting of a dipole and two or more additional elements, including a slightly longer reflector and a slightly shorter director. Electromagnetic coupling among the elements maximizes signal gain on both transmit and receive in the direction of the director.

YL (*CW abbrev.*) (*slang*) – 1. "Young lady," an unmarried female Ham.
2. Any unmarried female.

YL Radio League - An organization for licensed female Amateur Radio enthusiasts. Abbreviated as YLRL. Visit: **http://www.ylrl.org/** See: **YL, XYL**

YLRL – YL Radio League. See: **YL Radio League**

YR (*CW abbrev.*) – "Year."

Z

Zulu ("ZOO – lew")

Dah – dah – di – dit

Z – The letter used to represent electrical impedance.

zed - a phonetic pronunciation for the letter "Z" used to avoid operators mishearing the letter as "C" or other letter. It is commonly used on phone modes in the place of "Z," especially in Australia, Canada, New Zealand, and Great Britain. Example: An operator with the call sign W4ZHR might give his call as, "W 4 zed H R."

zero beat - Adjusting the frequency of a transmitter so it is on precisely the same frequency as another station.

zepp antenna – An antenna consisting of a single wire, one-half wavelength long on its design frequency, and fed from one end. This antenna was developed to be used as an antenna for zeppelins and other lighter-than-air craft from which it could be reeled out when in use and then reeled back in for landing or docking.

Z-signals – Three-character codes, similar to the Amateur Radio Q-code but beginning with the letter "Z." Used on CW, primarily in the military and in the Military Affiliate Radio System. Visit: **http://www.radiotelegraphy.net/zsignals.htm** See: **MARS, Military Affiliate Radio System, Q-code**

zulu (time) – Military term for Coordinated Universal Time. Means the same as Greenwich Mean Time and Coordinated Universal Time, which is the time at 0-degrees longitude, which passes through Greenwich, England. Represented by the letter "Z" after the time in 24-hour format. Example: 3:40 PM UTC is 1540Z. See: **Coordinated Universal Time, UTC, Greenwich Mean Time**

NUMBERS and PUNCTUATION

Ø – The number zero with a slash through it in order to distinguish a zero from the capital letter "O." (To create this character on most keyboards, hold down the Alt key while pressing Ø 2 1 6 on the numeric keypad.)

5/8 wave (*slang*) – An antenna that is 5/8 of a wavelength long. Usually applies to a VHF or UFH vertical radiator with a loading coil at its base.

1-by-1 call sign – A special temporary call sign consisting of a single letter, a number, and another single letter, issued to special event stations or for other distinct purposes. Example: The call sign N9N was issued for the special event station commemorating the fiftieth anniversary of the journey to the North Pole by the submarine USS *Nautilus*. Visit: **http://www.1x1callsigns.org/**

10 code - A series of abbreviations created originally for public safety communications but that were adopted and modified considerably by Citizens Band operators. The 10 code is not in use by Amateur Radio and is frowned upon when used by newcomers.

10 – 10 – Ten Ten International. See: **Ten Ten International**

11 meters (*slang*) – Another name for the Citizens Band. 11 meters is the wavelength of the 27 megahertz range assigned to that service. This was once a Ham band but was taken away to create the new CB service. Some resentment lingers.

33 (*CW abbrev.*) (*slang*) – "Love sealed with friendship and mutual respect between one young lady (YL) Amateur Radio operator and another." Adapted officially by the YLRL. See: YL, **YL Radio League**, XYL, **YLRL**

5-2 (*slang*) – The national 2-meter simplex calling frequency of 146.52. Example: "I'll be listening for you on 5-2." See: **calling frequency, simplex**

73 (*CW abbrev.*) (*slang*) – "Best regards." Though intended for CW use, "73" is used extensively on voice and digital modes.

73 *Magazine* – A former Amateur Radio magazine, now defunct.

88 (*CW abbrev.*) (*slang*) – "Love and kisses."

92 code - An early numerical code adopted for radio telegraphers. "88" and "73" were originally from that code but none of the other numbers remain in use in Amateur Radio. See: **73, 88**

807 (*slang*) - A beer. Named for a popular transmitting tube of the past that resembled an upside-down beer bottle.

9913 - - Common type of coaxial cable used by Hams as feedlines for antenna systems. See: **LMR-200, LMR-400, RG-6, RG-8, RG-8X, RG-58, RG-59**

? (*CW abbrev.*) – "Repeat."

/ (*CW abbrev.*) - "Slash" or "front slash." In Amateur Radio, typically sent before a designation that the station is operating portable or maritime mobile. Sent as dah-di-di-dah-dit.

ABOUT THE AUTHOR

Don Keith is an award-winning broadcaster, a best-selling author, and has been a licensed amateur radio operator since 1961. He was first licensed as WN4BDW in 1961, earned his General class license later that year, and then became an Amateur Extra class licensee in the mid-1970s, changing the call sign to N4KC.

He was twice named *Billboard Magazine*'s "Broadcast Personality of the Year," won every major broadcast journalism award in his state from the Associated Press and United Press International, and was an on-air personality, journalist, station owner, program director, and manager in a broadcasting career that spanned over two decades.

Don published his first novel, *The Forever Season*, in 1995. It has remained in print continuously since and was named "Fiction of the Year" by the Alabama Library Association. His more than two dozen other published works, fiction and non-fiction, cover such topics as NASCAR racing, broadcasting, college sports, submarines, biography, and World War II history.

Don lives in Indian Springs, Alabama, with his wife, Charlene, has three grown children, and three grandchildren. He operates all the shortwave amateur radio bands as well as VHF and uses most modes, including CW, SSB, PSK31 and FM. He enjoys DXing, contesting, antenna experimenting, and just plain rag-chewing.

Don's author web site is **www.donkeith.com**. His Amateur Radio web site is **www.n4kc.com** and features a number of stories and articles about the hobby and of interest to Hams.

Enjoy this book? You would certainly like Don's other Ham Radio books, "Riding the Shortwaves: Exploring the Magic of Amateur Radio" and "Get on the Air...NOW!" See more and read a sample at **http://www.donkeith.com/**

www.ingramcontent.com/pod-product-compliance
Lightning Source LLC
Chambersburg PA
CBHW020743180526
45163CB00001B/330